THE GRAND WESTERN CANAL

INLAND WATERWAYS HISTORIES

Edited by Charles Hadfield

The Ballinamore & Ballyconnell Canal. By Patrick Flanagan
The Birmingham Canal Navigations, Vol I. By S. R. Broadbridge
The Bude Canal. By Helen Harris and Monica Ellis
The Dorset & Somerset Canal. By Kenneth R. Clew
The Grand Canal of Ireland. By Ruth Delany
The Grand Junction Canal. By Alan H. Faulkner
The Grand Western Canal. By Helen Harris
The Great Ouse. By Dorothy Summers
The Kennet & Avon Canal. By Kenneth R. Clew
The Leicester Line. By Philip Stevens
London's Lost Route to Basingstoke. By P. A. L. Vine
London's Lost Route to the Sea. By P. A. L. Vine
The Nutbrook Canal. By Peter Stevenson
The Royal Military Canal. By P. A. L. Vine
The Somersetshire Coal Canal and Railways. By Kenneth R. Clew
The Thames & Severn Canal. By Humphrey Household
The Yorkshire Ouse. By Baron F. Duckham

in preparation

The Derby Canal. By Peter Stevenson
The Exeter Canal. By Kenneth R. Clew
The Forth & Clyde Canal. By Graham Matheson and D. Light
The Leicester Navigation. By Philip Stevens
The Nene. By Ronald Russell
The Oxford Canal. By Hugh Compton
The Shropshire Union Canals. By H. Robinson
The Stroudwater Navigation. By Michael Handford
The Warwick Canals. By Alan H. Faulkner

The canal near Tiverton about 1833, from a painting by D. C. Weatherley

THE GRAND WESTERN CANAL

by

HELEN HARRIS

With plates and text illustrations including maps

and a foreword by

CHARLES HADFIELD

DAVID & CHARLES : NEWTON ABBOT

0 7153 6254 2

Set in 11 pt Garamond, 2 pt leaded
and printed in Great Britain
by Ebenezer Baylis & Son Limited The Trinity Press Worcester and London
for David & Charles (Holdings) Limited
South Devon House Newton Abbot Devon

Contents

List of Illustrations

PLATES

Photographs not acknowledged above are from the author's collection

MAPS AND ILLUSTRATIONS IN TEXT

Foreword

ACROSS east Devon, from the Somerset border near Burlescombe to Tiverton, runs perhaps the most extraordinary canal in England. Go to Tiverton, and, clinging to the side of a hill outside the town, you will find its terminal basin, its outer walls incorporating massive former lime-kilns. Take the Sampford Peverell road, and you will cross it each side of Halberton and again in Sampford; and so you may follow it to an abrupt end in a cutting at Lowdwells beyond the Canonsleigh quarries.

To one's astonishment, this is no narrow or tub-boat canal, but one built for good-sized barges. Broad, once deep, with solid bridges and culverts, the Grand Western swings through the red earth of the Devon countryside as if it belonged there. The name itself, with its reminiscence of the later 'Great Western', gives a clue to its origin. For this eleven miles of watery beauty, now owned by the Devon County Council, represents an effort to save lives, ships and time lost rounding Land's End, by bringing south Wales coal landed at Somerset ports across Devon to Exeter. More, as a first instalment of a canal from the Exe to the Kennet & Avon, to give a through inland waterway route between London and Exeter.

Twenty-odd years after the first section was built, another on a smaller scale was added to extend the line from Lowdwells to Taunton to join the Bridgwater & Taunton Canal. This long-abandoned addition has special interest because of its inclined plane and seven vertical lifts. British engineers invented the canal lift, but though many prototypes were built, only these on the Grand Western worked in service until Anderton was opened in 1875.

Helen Harris has written a most interesting account of this canal venture. Speculation, over-optimism, bad forecasting and inadequate engineering control have bequeathed us the Grand Western Canal. A man must be hard to please who does not enjoy a spring walk along its towpath: hard to please, too, who is not interested in the remains of lifts, incline, aqueducts and wharves on the disused stretch to Taunton.

After reading Helen Harris's book, I feel myself a part of history. For most of 1924 I lived at Tiverton, and when not demonstrating my military incompetence in Blundell's O.T.C., I explored Tiverton wharf and the towpath towards Halberton. From late 1924 to 1927 I lived in Halberton village, and then for a year in the now ruined Rock Cottage, beside the Grand Western Canal below Rock House. Bicycling to school, I passed Elworthy's stone wharf, and saw the last of the canal's stone boats. At Rock Cottage the occasional horse-drawn maintenance craft with its sharp-pointed bow and square stern would pass our windows and disappear under Rock bridge, and most summer evenings I would walk my dog from Rock or Greenway bridges round the Swan's Neck.

I must have been fifteen when my father's solicitor gave me access to a tin box of old canal records still in his office, for his firm had been that of Mr Partridge, mentioned by Mrs Harris. Excited by my first piece of original research, I transcribed most of them. From that find of papers, from my walks round the Swan's Neck, my explorations past Sampford Peverell to Lowdwells, and my later discovery of the canal's remains towards Taunton, came my own enduring interest in waterways.

I commend this book to its readers, and the canal also. May both flourish together.

CHARLES HADFIELD

CHAPTER 1

Early Plans

THE Grand Western Canal was born out of man's passion, in the hazardous days of sail, for providing cross-country navigable links between the seas, and was conceived as part of a grandiose scheme which, if fully completed, would have connected the English and Bristol Channels by inland waterway. Throughout the eighteenth century and even into the nineteenth the idea for a water route across the south-west peninsula of England was constantly being raised in varied forms, recurring with a persistency almost comparable to that of the south-westerly winds which, alike in turn with those from north and south, whip up tumultuous seas against rocky coasts before shedding their rain on the green land of Wessex.

The south-west had seen its earliest canal building before the seventeenth century. In 1566 the first Exeter Canal—small in its original form (for craft of up to 16 tons only), and the first canal to be built in Britain since the days of the Romans—was constructed alongside the River Exe from below Countess Wear to Exeter, re-establishing the Devon city's navigable connection with the English Channel, of which it had been deprived since the building of obstructive weirs across the river some centuries previously. Then in 1638, in Somerset, the River Tone was made navigable for eight miles upstream from its confluence with the Parrett, and by 1717, by the construction of locks, for the whole way to Taunton, thus providing the county town with an inland waterway from the Bristol

Channel and the port of Bridgwater. With these two ingressions, one from the south coast and one from the north, it was perhaps not surprising that, by the mid eighteenth century and increasingly during the climate of canal mania of the early 1790s, ideas for cutting an inter-channel link, which would save vessels a stormy passage around Land's End and also provide water transport to inland towns and villages, should appear as an idealistically irresistible dream. Such ideas were formulated in numerous projected schemes, which included, in Devon and Cornwall, the Bude Canal's original plan for an extensive route from Bude bay to the navigable River Tamar, and the plan for the Public Devonshire Canal, which was intended to connect the Exe estuary and that of the River Taw in north Devon but which never came into being.

THE PROJECT INITIATED

An early move was one made by a group of Taunton men in 1768 when it was reported: 'a subscription of 10/6 afoot to set Mr Brindley to survey for a canal from Taunton to Exon.'[1] The survey was carried out in 1769 by Robert Whitworth, working under James Brindley's supervision, and the line designed to run either from Topsham on the Exe estuary, five miles downstream from Exeter, to Cullompton, or from Exeter itself to Cullompton or Tiverton, and thence to Wellington and Taunton. The barge route would be extended beyond Taunton to the Bristol Channel by means of the Tone Navigation as far as Burrow Bridge, and then by a new canal through Bridgwater, Glastonbury, Wells and Axbridge, to Uphill at the mouth of the Axe, just south of Weston-super-Mare. But no action was apparently taken to implement Whitworth's scheme.

Nearly a quarter of a century later the idea was revived, and on 1 October 1792 a meeting was held at the Half Moon Inn, Cullompton, to consider the making of a navigable canal from Taunton to Topsham. Francis Colman accepted the chair, and the others present were: Francis Rose Drew, Thomas Newte, Rich Hall Clarke, William Lewis, Nicholas Dennys, Richard Lardner, Nicholas Were, the Rev Edward Drewe, the Rev William Walker, the Rev John Brutton, Henry Skinner, William Brown, Martin Dunsford, John Fowler, Henry Dunsford, Thomas Wood and Thomas Pannell. Henry Brutton, attorney at Cullompton, was appointed secretary and solicitor. The meeting resolved unanimously:

> That the plan for making a navigable canal from Taunton to Topsham with a branch to Tiverton is clearly practicable at a very moderate expense, and that it will be productive of the very greatest advantage to Trade, Commerce and Agriculture.
>
> That among the advantages that may arise from making this canal, appears to be the easy communication betwixt the Bristol and the British Channel, instead of sailing round the Land's End, which requires various winds, and both in winter and war is a tedious and dangerous navigation.
>
> That this canal will afford a safe, easy and expeditious conveyance for coals, lime, timber, (and other materials for building), iron, cheese, salt, groceries, hardware, wool, etc.
>
> That it is an object of national importance by the ready conveyance of timber to His Majesty's dockyards from the north of Devon, the counties of Somerset, Gloucester, and (particularly the Forest of Dean) Hereford, Worcester, etc.
>
> That the general communication between Ireland and Scotland with the western part of England and the Bristol Channel would also be rendered much more easy and safe.

A second meeting was planned for the 29th of the month at which the proprietors of land, merchants, manufacturers, and 'all other persons willing to promote so laudable an undertaking' were desired to attend and meanwhile requested to collect all information in respect of tonnage of the various articles enumerated, as carried from town to town, conveyed coastwise, or exported.[2]

As a result of the subsequent meeting, and typically of the times, the alternative advice of a different engineer was sought, and John Longbotham was called in. The Yorkshire-born Longbotham was a man of thirty years' engineering experience;

earlier he had surveyed a route for a canal across the Pennines from Liverpool to Leeds and during the 1770s had been engineer for its construction to the resurveyed plan of Brindley and Whitworth. In January 1793, upon Longbotham's attendance at a meeting, the committee unanimously resolved that he should immediately proceed to survey the line for the intended canal and present a report.[3] This was carried out forthwith and the cost of the scheme for the canal from Taunton to Topsham, with lateral lines to Wellington, Tiverton and Cullompton, estimated by Longbotham as £166,724, of which £22,229 was for the branches.[4]

In June, when the plan and estimate had been received, it was reported that the committee had ordered a revision of the plan to be made by Robert Mylne (engineer of the Gloucester & Berkeley Canal), and at the same time it announced that a further subscription would be necessary and that books would be opened at the next general meeting for such subscribers as were willing to enlarge their shares.[5] At this meeting, held on 7 August, the subscribers were apparently well pleased with the situation and it was resolved that the deficiency of the sum already subscribed to the complete estimate should be filled by taking a second set of shares at £100 each, according to original subscriptions. The books showed that, out of the 90 original subscribers, 88 came forward at the meeting to enter a further total subscription of £71,200.[6]

JESSOP'S REPORT

Anxious to have the best possible engineering opinion of the day, the committee in the same year engaged William Jessop to advise on the comparative suitability of the drawn-up schemes, the possibility of further improvements, the right dimensions and best terminal points and whether potential business would justify the outlay. Apparently assisted by Hugh Henshall, brother-in-law of James Brindley, Jessop conducted his appraisal and presented his report[7] at a meeting of the select committee held at the Half Moon Inn, Cullompton on 29 November 1793, with the Rt Hon Sir George Yonge, Baronet and Knight of the Bath, in the chair.

Jessop found that the cost of the canal, including the branches, would be lower by Longbotham's plan than by Whitworth's, and that the length of the main line would be about 2½ miles less. He considered that Longbotham's line in general need not be materially altered, but made a few suggestions for improvement and shortening. He had doubts about the proposed branches to Tiverton and Cullompton, which he suggested should be omitted, and a branch to Wellington he also felt would cost too much for so short a distance; however, the expense of a short branch planned to run to the limestone rocks at Canonsleigh—site of important quarries—though not absolutely necessary, he considered 'too trivial to weigh objection'.

Regarding another branch, designed to run to the Exeter–Honiton road near Sowton for trade with Exeter, and destined soon for reconsideration for the part it played in a brooding matter of local contention, Jessop wrote: 'The Branch proposed to the Honiton Road will certainly be advisable, and is the nearest Approach that can with Propriety be made towards the City of Exeter, and will be productive.' In addition to the feeders planned for the canal he also suggested that a reservoir might be made 'somewhere above Culmstock' to provide compensation water for mills at times when rivers were running low.

Jessop considered, as the canal was intended to link with the seas, that it should be large enough to accommodate barges carrying up to 50 tons, 14ft 6in or 15ft wide, which could navigate the Exe estuary from Topsham to Exmouth and cross the Bristol Channel carrying coal from south Wales. He advocated a waterway with a width of 42ft instead of 33ft as previously suggested, except in deep cuttings and on embankments, with a depth of 5ft, and recommended locks of 8ft rise rather than 5ft. These he stated would cost less overall, would enable boats to pass more quickly, be less likely to be out of repair and less liable to damage by boats striking against the breast of the lock.

Regarding the canal's potential business, Jessop stressed the importance of the coal trade, pointing out that for success it was essential that coal should be obtained and conveyed at a suitably economic price for supplying a part of Exeter's consumption. This would hinge on whether or not coal imported from Wales was subject to coastwise coal duty, which applied to coal shipped from west of Newport that passed west of the Holms, and so to coal carried on the Tone. Coal carried across the Bristol Channel from Newport that passed east of the Holms was free of the duty, so that it would be prudent to extend the canal beyond Taunton to the originally intended terminal point of Uphill—easy flat country for canal construction—so as to take advantage of this facility. The waterway might even go as far as Nailsea, where also there were collieries and to which there was a possibility of a canal link with Bristol. By this means coal obtained at the lower rate could be conveyed more cheaply than by land carriage and would avoid the sea charges involved for that brought to Exeter around the coasts. Jessop estimated that coal brought to Uphill for 9s (45p) a ton, carried from there the 71 miles by canal to the point on the Honiton road where the Exeter branch would end, at 1½d a mile— amounting to 8s 10½d (44½p) —and transported by land the remaining distance into the city for 1s 6d (7½p), would be available in Exeter for 19s 4½d (96½p) a ton, whereas Newcastle and Liverpool coal was currently selling there for 24s (£1.20). Besides Exeter's domestic demands, much small coal was used there for lime-burning and brickmaking, and it was also required by towns along the projected line of the canal and by the great limeworks at Canonsleigh near Burlescombe.

At that time 20,000 tons of lime a year were being sold from Canonsleigh for application to the land, some of it carried 25 miles from the works. It was the principal manure along the

Page 17 (*above*) The Tiverton basin. Formerly there were wharves on either side, with the tops of limekilns in the area between the two sheds on the right; (*below*) some of the limekilns below the Tiverton basin; they were fed directly from the wharf above, but their bowls have long been filled in.

Page 18 (*above*) Manley bridge, stone-built, between Tiverton and Halberton; (*below*) a typical accommodation bridge north of Halberton village. Metal now replaces the former wooden structure.

whole line of the planned canal from Burlescombe to Topsham and reports indicated that on about two-thirds of the tillage land 2½ tons of lime were applied per acre every 6 years. If the price was lower more was expected to be used. Limestone could be obtained at Canonsleigh for 6d (2½p) a ton, but at Topsham cost 2s 6d (12½p); if coal were obtainable at a lower price lime burnt more cheaply could be carried to Topsham and Exeter and used in much greater quantities. Jessop estimated that, to supply land carriage as well as that of the canal, 40,000 tons would be burnt at Canonsleigh, needing 10,000 tons of small coal, delivered at the works for 12s 5d (62p) a ton. Besides Canonsleigh, there was also another limeworks on the line of the canal, at Winsheer near Wellington.

A further important form of trade envisaged by Jessop was the transport of flat paving stones obtained in Somerset and the neighbourhood of Bristol; the latter were considered the 'best in the kingdom' and would be particularly accessible if a projected canal onwards to Gloucester, which would pass through considerable quarries of the material, were made.

Jessop estimated that the carriage of coal and lime would produce a revenue of £8,311. He stated that in similar projects he had generally found that where, on accurate information, it appeared at the outset that articles conveyed would produce a 3 per cent revenue on the expense, that the increase of carriage brought about by a reduction in its cost made the undertaking practicable. In conclusion, however, he re-emphasised that the scheme's success depended on its extension from Taunton towards Bristol. Two rival groups were currently preparing to do just this, one supporting a proposed Bristol & Western Canal from the Avon at Morgan's Pill near Bristol to Taunton, and the other a consortium set on reviving a section of Whitworth's scheme as the Taunton & Uphill; each was anxious to have the support of the Grand Western and both had at this time also sought Jessop's advice. A third group proposed a canal across Somerset from the Bristol Channel near the collieries at Nailsea by Bridgwater and Chard to the English

Channel at Seaton: this would cut across the others and so, everyone hoped, generate further traffic.

At the following general meeting of subscribers held on 16 December, it was resolved 'that the plan and survey of Mr Longbotham of the intended canal, as corrected by Mr Jessop, be approved of, and the same be carried into execution.' A new select committee was appointed, consisting of: Sir George Yonge (honorary and corresponding member), Francis Rose Drewe, Francis Colman, Richard Graves, Rich Hall-Clarke, John Wedgwood, John Chave, Samuel Bansill, John Fortescue, James White, Joseph Sanden, Edmund Granger, John Cole, George Waymouth, Nicholas Were, Robert Baker, William Lewis, Thomas Heathfield, Martin Dunsford, the Rev Brutton, the Rev Follett, John Fowler and Mark Weston. The thanks of the meeting were given to Sir George Yonge 'for the great trouble he has taken, and for his very able conduct in the chair' and also to the former select committee 'for the zeal and constant attention to the business of the undertaking.'

RENNIE'S SURVEY

But in 1794, perhaps because of doubts amongst the subscribers, possibly on account of certain rumblings of opposition coming from the direction of the city of Exeter, it was felt desirable to obtain the advice of yet another engineer, and John Rennie, then engineer of the Kennet & Avon Canal, was called in to give his opinion and make an alternative survey.

The plan produced by Rennie following his survey of 1794[8] showed the proposed canal running in a generally north-easterly direction for about the first two-thirds of its way, from the Exe estuary as far as Burlescombe, and then taking a course roughly due eastwards to Taunton. From Topsham the route first followed the west side of the valley of the River Clyst, but after running through Sowton crossed to the east of the river between Clyst Honiton and Broadclyst and crossed back again before reaching the parishes of Clyst St Lawrence and Clyst

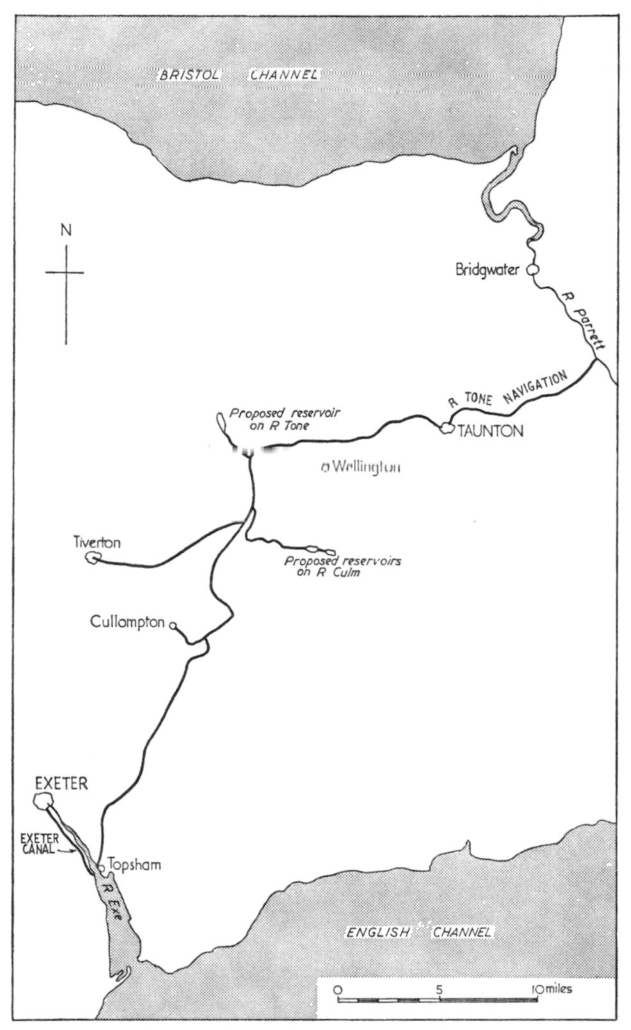

FIGURE 1. The approximate route of the proposed Grand Western Canal, from Topsham to Taunton, with branches to Tiverton and Cullompton, as planned by John Rennie in 1794

Hydon. After passing into the Culm valley a branch led off westwards to Cullompton, the main trunk continuing through Willand, west of Uffculme and across Leonard Moor, where another westward branch left to run through Sampford

Peverell and Halberton to Tiverton. Shortly past the junction with this branch a feeder was shown joining the canal from the east, with two reservoirs sited beyond Culmstock, on the River Culm between Hemyock and Clayhidon. A further feeder was to supplement the canal on its continuation through Burlescombe and Holcombe Rogus, bringing water from a reservoir to be situated on the River Tone a short distance to the north. Past Burlescombe the route was shown entering the valley of the River Tone north of Thorne St Margaret, and, running north of Wellington and south of Nynehead, keeping in fairly close proximity to the River Tone for the remainder of its way to Taunton, where it joined the river about half a mile above the bridge in the town. Rennie had reinstated the branches to Tiverton and Cullompton, disliked by Jessop, but had omitted the one at Sowton, near Exeter, which Jessop had advocated. (For lengths, rises and falls see Appendix 1).

John Rennie communicated with the committee regarding mills in a letter dated 2 July 1795.[9] He was of the opinion that two reservoirs with a joint capacity of 66,380,000 cu ft of water would supply the canal completely for 140 days without assistance. He wrote: 'I am of the opinion the canal may be amply supplied with water by their means without injuring the Mills in any degree, supposing the whole of the River in ordinary times necessary for their supply'. But he felt that the mills nearer Exeter might not be so well able to spare water as those Jessop had considered farther up the Culm and he advised taking powers in the Bill either to take water wholly or partially from the mills on the Culm, making proper compensation for injury sustained, or to erect reservoirs for the necessary supply, taking surplus water from the Culm in flood times only.

At a general meeting of subscribers held on 13 July 1795 under the chairmanship of Richard Graves, at the usual Cullompton meeting place, it was resolved: 'That the resolution relating to the Line of Canal, as proposed by Mr Longbotham, and approved of by the General Meeting of the 16th

December 1793 be now rescinded; and that the line, as altered, improved and amended by Mr Rennie, now produced, be adopted.' It was also resolved by a great many of the subscribers present to bring in a Bill during the next Parliamentary session to carry the plan into execution. The secretary was instructed to give the necessary public notices and John Rennie engaged to help with the Bill.[10]

EXETER'S OPPOSITION

By this time much anxiety had been aroused amongst the citizens of Exeter[11] at the prospect of the Grand Western Canal, and voices were being raised. Their main concern was the apparent threat to the trade on the city's own canal from Topsham. The Exeter Canal, originally constructed under powers granted by an Act of Henry VIII's reign, had since been expensively improved from time to time. In the 1670s dredging and improvements were carried out and between 1699 and 1701 the canal was enlarged and straightened so that it could carry not only barges and lighters but also ships and vessels of up to 200 tons, thus providing a valuable communication from sea to city, greatly to the advantage and benefit of Exeter's trade and commerce.

The cost of the work had, however, been heavy; £21,000 had been expended on it during the last 25 years of the seventeenth century, and when a Bill was sought for raising a further £5,000 for completion there was opposition, as a result of which the then Mayor and Chamber of the City of Exeter did not press their application but completed the work—amounting to a sum far higher than the estimate—with their own money. A considerable part of this debt was still outstanding, on a mortgage of the canal and other property, and the interest on this, in addition to annual maintenance, which had amounted to over £33,000 in the 37 years from 1758, meant that the current Mayor and Chamber were only receiving a very small return on the original investment. It was felt, not unjustifiably, that the Exeter Canal, established at such great expense, ought not to be put in hazard by any new speculation such as that of the canal now proposed from Topsham to Taunton, openly designed as it was for supplying part of the city's coal requirements. The resulting fall in revenue from the carriage of coals on the Exeter Canal, which formed two-fifths of its current income, might make it impossible to maintain it for the bigger ships and vessels so necessary to the city's trade.

The second point in Exeter's disquiet concerned the city's supply of water from the River Exe, which it was feared would be impaired by taking water from the River Culm to feed the proposed canal. Joining the Exe near Stoke Canon, about four miles north-east of Exeter, the Culm contributes a fair share of the Exe's water below this point and was said to be principally depended upon in dry seasons. It was stated that half of the water flowing in the River Exe between Exeter and Topsham was consumed in maintaining the Exeter Canal fit for heavy ships, and in addition there were the city's domestic and industrial requirements to be considered. Much of Exeter was supplied with piped water from the Exe lifted by means of an 'engine' erected some time previously, for which for nine months of the year the existing flow was no more than sufficient, and at times scanty or totally inadequate. There were also many fulling mills necessary for carrying on the woollen manufacture of the city, as well as corn mills, situated on the Exe, and in recent years a large mill for spinning cotton had been established adjoining Trew's weir. It was contended that any diminution of the water in the River Exe would be highly injurious and prejudicial to the Exeter Canal—perhaps a means of its total destruction—and would also prevent the necessary supply of water for working the water engine and the mills.

A yet further reason for unease lay in the fact that under ancient charters and grants large rents were payable from Exeter to the Crown, in return for which the city was entitled to receive various petty customs or town dues on all goods imported into the port of Exeter, covering in extent the River

Exe to the sea and the sea coast from Lyme Regis to Teignmouth. The city was also entitled to dues on all goods landed on or shipped from Exeter quay which it also possessed, and at the ancient quay at Topsham, rented from the Crown. The latter extended to the limits of the manor of Topsham within which the intended canal was proposed to run, although no quay or wharf could legally be erected within these limits except by the city authorities or with their consent. The proposed Grand Western Canal from Topsham to Taunton was seen by the citizens of Exeter, therefore, as a potential danger if not a possible cause of actual ruin to its own Topsham–Exeter canal and to the mills, and also as a threat to its quay and port dues.

The canal committee gave due consideration to the city's opposition. On 5 December 1795 Henry Brutton, its secretary and solicitor, wrote to the mayor, mentioning recent consultations with the engineers Wedge and Easton, acting under Rennie, on the probability of supplying water by reservoirs at Ashbrittle instead of by a feeder from Culmstock and stating, somewhat bravely, that it would be in the interest of the proprietors to carry coal on to Topsham as they would thus gain more tolls. On 23 February 1796, the day before the second reading of the Bill, against which the Mayor and Chamber of the City of Exeter were petitioning, Sir George Yonge wrote that the committee was willing to guarantee the city against loss of dues. The canal's financial prospects were becoming less promising than they had seemed earlier.

THE ACT OF 1796

On 24 March 1796 'An Act for making a Navigable Canal from the River Exe near the Town of Topsham, in the County of Devon, to the River Tone, near the Town of Taunton, in the County of Somerset; and for cleansing and making navigable a certain Part of the said River Tone; and for making certain Cuts from the said Canal' was passed (36 Geo III c. 46).

The list of 158 subscribers included many local names and also those of some nationally-known figures (see Appendix 2).

By the Act, which ran to 119 clauses and referred to a 3-page schedule, the Company of Proprietors of the Grand Western Canal was empowered to purchase land and to make and keep navigable for 'Boats, Barges and other Vessels' a canal from the tideway or navigable channel of the River Exe at Topsham to the River Tone in the parish of Bishop's Hull, passing through the parishes of 'Topsham, Heavitree, Sowton, Pinhoe, Honiton's Clyst, Broad Clyst, Clyst St Lawrence, Clyst Hydon, Plymtree, Cullompton, Kentisbeer, Uffculme, Welland, Halberton, Burlescombe and Holcomb Rogus' in Devon, and in Somerset through 'Ashbrittle, Kitsford, Langford Budville, Thorn Saint Margaret, Wellington, Ninehead, Hill Farrants or Farrance, Bradford and Bishop's Hull.' Cuts from the canal were authorised to Cullompton, to Tiverton (from a point in Burlescombe parish and running through Sampford Peverell, Halberton and Tiverton parishes) and also to Wellington, the latter apparently resulting from a revival of the earlier suggestions, probably due to pressure from subscribers in the locality. There were to be two reservoirs in the valley of the River Culm in Devon, and two in the valley of the Tone in Somerset, with feeders from them connecting with the main canal at points in Burlescombe and Holcombe Rogus parishes respectively, and in addition to the use of water from the rivers concerned the company was also granted the right, with certain exceptions, to take water from any rivers, springs or streams within 2,000 yds of any part of the canal and cuts. The making of inclined planes, if necessary, was also authorised.

For the protection of the mills on the Rivers Tone and Clyst, the Act stipulated that the company was not empowered to take water from streams supplying such mills but only the surplus water of the rivers, taken into the reservoirs in times of excess; gauges or watermarks were to be fixed to regulate this, supervised by two engineers, one appointed by the owners of the mills and the other by the company, who would refer to an umpire in case of dispute. For the security of mills dependent on the water of the River Culm—those actually on it and others on the Exe—and the City of Exeter's interests, the company was only authorised to take the surplus or flood water into the reservoirs. The system and its necessary gauges were to be watched over by six commissioners, three appointed by the City of Exeter and three by the company, again with reference to an umpire if necessary. The Act fully safeguarded Exeter's water supply: extraction of any water from the Culm other than flood water or from streams feeding it, and also from any part of the Exe or its tributaries was prohibited, any offender being liable for a penalty of £50 payable to the city authorities.

The land which was to be taken, mineral rights of which were to continue to belong to original owners, was not to exceed 30yds in breadth, except at docks, basins or reservoirs, or where embankments or cuttings, passing places or wharves were required. There were the general requirements regarding the provision of substitute roads for any taken or cut through, clay puddling to prevent water oozing from the canal on to the land, fixing and maintenance of fences, and the provision of bridges, culverts and watering places for cattle where these were made necessary. The company was authorised to cleanse adjoining watercourses at landowners' expense and was required to repair any flood or accidental damage to its works which might cause damage to nearby land. An important clause was one which authorised the company to cleanse and make navigable part of the River Tone, for a distance of about 500yds from the point in Bishop's Hull parish where the canal was to end downstream to Taunton bridge, by which it was considered 'the said Canal would be rendered more beneficial to the Publick.' This length of river was declared to be a part of the canal and vested in the company, which was empowered to build warehouses, wharves, quays and landing places on the riverside.

The Act authorised the company to raise £220,000 in £100

shares to execute the undertaking, with no proprietor having less than one share nor more than 100, and each share carrying one vote which could be given by proxy. Provision was also made for the raising if necessary of an additional sum of £110,000, either by further calls amongst the proprietors themselves or the admission of new subscribers, or by borrowing on a mortgage of the undertaking. The Act directed that the first General Assembly of the Company of Proprietors should be held at Cullompton on the Thursday three weeks hence, when a 21-man committee was to be chosen and officers appointed, and outlined the proceedure under which it was to be run. The company was bound to admit to shares, if requested, owners of land amounting to 1 acre and over, or ⅛mile in length, through which the canal or its cuts was to pass (or reservoirs were to be situated), who would subscribe according to the quantity of their land purchased, at a rate not exceeding 1 share for every acre or ⅛mile. In the event of owners declining the privilege, their tenants might become subscribers under similar terms. The company was not required, however, to admit landowners and tenants on this basis for a greater sum collectively than £50,000.

The Act empowered the company to take tolls for goods carried on the system at rates fixed at any General Assembly, but not exceeding:

Lime, limestone and other articles used for manure	1d per ton per mile
Lime and limestone (except for the purpose of manure), ironstone, iron ore, lead ore and all other ores, stones, tiles, slates, bricks, flagstones, clay and sand, and all materials for repairing the roads	2d per ton per mile
Rough timber, iron in pigs and bars, lead in pigs and sheets, tin in lumps and bars, charcoal, salt, cheese, corn, hay and straw	3d per ton per mile

Coals, culm, coke and cinders	2d per ton per mile
Wrought metals, oils, wines, liquors, groceries, earthenware, and all other goods etc not specified	4d per ton per mile

A fraction of a mile, where involved, was to be deemed as one mile, but a fraction of a ton charged as a proportion of the rate. No vessels carrying hay, straw or corn in the straw, road repair materials or any kind of manure were to be allowed to use any locks unless water was flowing over the waste weir, except with special permission. The company was to fix the rate of wharfage payable in respect of any goods remaining on its wharves for over 24 hours, and also the price to be charged for the carriage of persons, cattle and small parcels.

On completion the canal was to be measured, and stones or posts erected at every mile. Each owner or master of a vessel (other than a pleasure boat) was required to have his name and place of abode, and the number of the boat, entered by the clerk to the company, and also his name and the boat's number correctly painted on its side, together with appropriate gauge marks indicating loading capacity. Owners were themselves answerable for any damage they or their servants, horses or boats caused to the undertaking or to property adjoining, and there were penalties for any unqualified person carrying or having on board fishing nets or other instruments or guns for taking and destroying fish or game.

Regulations also covered the provision of vessels' passing places, the passage of boats through locks, the removal of obstructions, and penalties for various misdemeanours, and contingency rules for any possible future private mining near the line of the canal were laid down. Lords of manors on the line were empowered to construct and use wharves on their wastes, and owners to do so on their own lands; if they did not, the company could. Fishing rights were reserved for the appropriate lords of manors and landowners, and owners of land and others were entitled to use pleasure boats and boats employed in husbandry free of charge under certain restrictions.

Full protection of the rights of Exeter Corporation was assured by the Act's later clauses. Under these it was ruled that coal, or culm, carried on the Grand Western Canal but intended for the city of Exeter, must pay Exeter Canal dues, whether it passed to the city by the Exeter Canal or by land carriage. Any landed ostensibly for local use, but then illicitly carried to Exeter would be liable to double dues. The city was also to receive the petty customs to which it was entitled on all coal, culm and other goods landed on its quays and those at Topsham, as well as dues for the use of the cranes.

The first general meeting was duly held on the following 21 April at the Half Moon Inn, Cullompton and the committee chosen; Henry Skinner was elected treasurer and Henry Brutton reaffirmed as secretary and solicitor to the Company of Proprietors.[12]

But, though the Grand Western had its Act, the Bristol & Western's Bill for a line from Morgan's Pill to Taunton had been defeated and dropped, and the Nailsea–Seaton scheme had faded away. In 1797 further consideration was given to extending the proposed route—important for viability—from Taunton to the north Somerset coast, and in July 1797 it was reported that the company was intending to petition for another Act for this purpose.[13] But by this time the country was fully embroiled in the war with France, and in the ensuing years of preoccupation, raised costs and reduced manpower, ideas for the canal passed into the shadows, and were to remain there for the next decade.

CHAPTER 2

The First Stage of Construction

REVIVAL OF INTEREST

By 1807 there was a revival of interest in the prospective Grand Western Canal and fresh moves were afoot towards getting it started. In 1807 and 1808 the Kennet & Avon Company, whose canal from the River Kennet at Newbury to Bath was under construction, was approached for financial support, but was too heavily committed to assist. On 30 June 1808 a resolution by Grand Western shareholders called on two of the subscribers, Sir George Yonge and the Rev Herman Drewe, to investigate the possibilities of commencement.[1] A further delay followed, mainly because of the difficulty of raising the necessary money. Many of the original subscribers had withdrawn or died since the Act was passed and only 1,297 of the 2,200 shares were paid up. But at the end of February 1810, when a meeting was called to receive the investigators' report, it was learned that a deputation from certain proprietors of the Kennet & Avon Canal had agreed to take over the 903 unappropriated shares and had made the necessary deposits.[2]

By this time Rennie's Kennet & Avon was nearing completion, and a project to extend it from Bath to Bristol was being canvassed, by a canal that would by-pass the Widcombe flight of locks and the Avon Navigation. Again, the old Bristol & Western scheme was revived as the Bristol & Western

Union. Both the Bath & Bristol and the Bristol & Western Union had Rennie as engineer. Together with the Grand Western, promoters envisaged a 200-mile inland waterway from London to Exeter on which 50-ton barges would travel without shifting cargoes.[3] Furthermore, the cross-Somerset canal scheme had been revived in October 1809, and in 1810 Rennie was surveying for this also. So the possibility of starting the Grand Western was considered against a heady background, and with an ambitious engineer.

On 12 April 1810 a special meeting of the Grand Western Company, chaired by Sir George Yonge, was held at Cullompton. With the originally estimated £220,000 now seeming unlikely to be enough for the work, the meeting was to consider the raising of another £110,000, and, amongst other matters, whether to begin construction.[4] But the meeting took a wider and altogether more far-seeing decision. Perhaps as part of a bargain, Sir George Yonge had promised his Kennet & Avon friends that he would use his influence to press for the proposed Bristol & Western Union scheme (later called the Bristol & Taunton—precursor of the eventual Bridgwater & Taunton Canal). In any case, the desirability of a continuation from the Grand Western to the Bristol Channel had already been acknowledged. Whatever lay immediately behind it, a close link with the Bristol & Taunton was initiated with the meeting's resolution that a subscription should be immediately opened for defraying the cost of a canal through Somerset to Bristol, not exceeding £330,000. One third of the subscription was to be offered to the proprietors of the Grand Western Canal, one third to the proprietors of the Kennet & Avon, and the remainder to the owners of land through which the canal would pass, the inhabitants of the principal towns in the county of Somerset and to the port of Bristol.

John Rennie was appointed engineer of the Grand Western, Isaac Cooke of Bristol solicitor and Messrs Stuckey & Co (bankers at Bridgwater, Langport and Bristol) treasurers. A deposit of £2 per share was to be paid to the treasurer by the coming 1 July, £2,000 of which was to be retained for defraying current expenses and the surplus invested in Exchequer Bills until required. Twenty-four committee members were named and requested to undertake management of the subscription, to apply to the landowners for permission to survey the line of the canal, and to take whatever other measures might be necessary.[5]

A fortnight later Sir George Yonge wrote from Hayne House, near Cullompton, to Richard Hall Clarke, who had been absent from the meeting:

> At a meeting of the Proprietors of the Grand Western Canal held at Collumpton on the 12th Inst to determine on the measures necessary to be adopted for the immediate prosecution of that object, the Gentlemen present considered that to give full effect to the Undertaking as well as to promote the accommodation of the Public in general, a navigable Communication should be continued from that Canal through the County of Somerset to the Port of Bristol; by which, and the extension of the Kennet & Avon Canal to Bristol, an Inland Navigable Communication would be made from London all through by Bristol to the Port of Exeter. A Subscription was immediately opened and £130,000 subscribed in the Room, and You Sir, were named one of the Committee to manage it. I therefore have the honour to transmit to you a copy of the Resolutions and to request the favor of your Aid and Assistance to the Undertaking which cannot fail to be of great National as well as local Utility.[6]

And on 19 April the following notice appeared in the *Exeter Flying Post*:

> Books opened by Henry Skinner of Cullompton and Henry Stokes of Shorter's Court, London, where such proprietors as are desirous of taking their proportion of such additional subscription, must apply on or before the 5th May
>
> Henry Brutton
> Thomas Merriman* Secretaries.

* Thomas Merriman was the clerk of the Kennet & Avon Canal Company

WORK IS BEGUN

The other important resolution passed at the meeting of 12 April was that work on the canal should commence, and this was very promptly begun on the following Monday, 16 April 1810. The *Exeter Flying Post* of 19 April reported the occasion as follows:

> Monday last, pursuant to a resolution of the general meeting of proprietors, held the 12th instant at Cullompton, the great work of the Grand Western Canal was commenced on the summit level in the parish of Holcombe, on land belonging to Peter Bluett Esq., for which occasion the first turf was cut with all due form and ceremony by the Rt Hon Sir George Yonge, Bart, chairman of the meeting, assisted by the lady of John Brown, of Canonsleigh, Esq., who attended for that purpose, in the presence of a numerous body of spectators of all ranks, who testified their joy at the commencement of the work which promises the greatest benefits to the whole country, not without the hope and prospect of its being the source of still further advantages and improvements. The day being fine, added to the pleasures of the scene. Money and cider were distributed to the populace, while the liberal hospitality of Holcombe Court, to the genteeler sort, closed the scene in a manner suitable to the occasion, and worthy the owners of that respectable mansion.

It perhaps seemed odd to commence construction of a canal in the middle part of its route rather than at one end. But the decision to cut first these 2½ miles of the main line and the 9-mile branch to Tiverton—the section anticipated as the most difficult and expensive—to a suitable width and depth for carrying large barges, was made in order to take early advantage of the large potential trade in lime and limestone envisaged between the Canonsleigh quarries and Tiverton and the agricultural area between.

Work proceeded under the eye of the superintendent, John Thomas. Thomas had superintended the building of much of the Kennet & Avon and was at this time still its manager.

Page 35 (*above*) The aqueduct which carries the canal across the former Tiverton Junction-Tiverton branch railway; (*below*) the high embankment on which the canal is carried north-east of Halberton village.

Page 36 (*above*) Twin culverts which carry a stream through the canal embankment near Burlescombe; (*below*) south portal of Waytown tunnel by which the canal passes deep beneath the Wellington–Holcombe Rogus road. Remains of a haulage chain can be seen on the left.

Doubtless on Rennie's advice, it was now decided to cut the summit level 16ft lower than had been first planned and to do without locks on the branch, thus saving water, a decision which contributed to the need for a further Act, in 1811, and which very much increased the cost of excavation and construction.[7] During the work various springs of water were found, particularly towards the level's eastern end, which bubbled up in considerable quantities through the canal bed, the cutting of which had apparently intercepted the source of the well-known Lowdwells spring which shortly afterwards ceased to flow.[8] These, together with some adjacent minor streams, provided the summit level with its relatively static water requirements.

The years which immediately followed the inception of the canal's construction in 1810 were ones which brought a lasting change to this east Devon landscape, with its rolling hills, its pastures and arable land divided by long-established substantial hedgebanks, and ancient farmsteads and quiet villages embodying the agricultural scene. The entire population of the four rural parishes through which the summit level's route ran—Holcombe Rogus, Burlescombe, Sampford Peverell and Halberton—amounted to no more than 3½ to 4 thousand, and that of Tiverton at the western extremity; the only sizeable town in the locality, was less than double that number. Men at work at the Canonsleigh quarry face and those who blinked away the acrid smoke of its limekilns could look across and see, taking shape, the new means of transport soon to be used for carrying away the produce of their labours, for which, it was said, there would be an increasing demand; and many who encountered the work in progress as they made their way with horses and carts to the quarry for supplies of stone and lime realised that their journeys would soon be unnecessary.

As the canal excavation which gashed open the land's surface gradually became extended, and the rich red mud was carried afield, a new element figured in the lives of the native population—the navvies who came in their scores from all parts of the

3

kingdom to descend upon the district and manhandle the construction with picks and shovels; tough, hard men, accustomed to rough itinerant living, determined in demanding their due rewards and often, in their transient existence, out for what they could get, regardless of the consequences. In April 1811, at the time of the annual Sampford Peverell cattle fair, there was a navvies' riot. A number of men were discontented at the time due to a delay in payment of their wages, which they had sought at both Tiverton and Wellington. On the Monday of the fair, after indulging in considerable drinking and committing 'various excesses' at Tiverton and other places, the navvies congregated at Sampford Peverell where they took advantage of the occasion to give full vent to their feelings, and made themselves a general nuisance. In the evening, after much rioting, 300 or more of the men had assembled in the village. An object of their disfavour was a Mr Chave, who, it appeared, had not long since debunked the story of a recently alleged local 'ghost'; they met and recognised him on the road to the village and followed him to his house, where they broke the windows. Fearful of further violence, Mr Chave apparently thought it necessary for his defence to fire a loaded pistol at his assailants, with the disastrous result that a man named Helps was killed outright. A pistol was also fired by some other person in the house, seriously wounding another of the navvies, and a carter employed by Chave was severely beaten by the mob, which was threatening to pull down the house. At the subsequent inquest, held at Sampford Peverell, the jury returned a verdict of 'justifiable homicide'. Controversy and ill feeling, however, continued.[9]

THE ACT OF 1811

In 1810 it had been realised that the route of the Tiverton branch as laid out in the Act of 1796 was not wholly practicable, one aspect being that if built according to the plan the canal would run directly through the village of Halberton.

Accordingly on 13 September that year notice was given of an intention to vary the line, and on 15 June 1811 'An Act to vary and alter the Line of a Cut authorised to be made by an Act of the Thirty-sixth Year of His present Majesty, for making a Canal from the River Exe, near Topsham in the County of Devon, to the River Tone, near Taunton in the County of Somerset; and to amend the said Act' was obtained. (51 Geo III c. 168.)

Under the Act power was given for the cut of the Tiverton branch, which was originally authorised to be made from a field in Burlescombe parish owned by Richard Corner to one at Tiverton belonging to Robert Hawks, to be varied to a route which, it was stated, would be more beneficial to the company and more convenient and useful to the public. The new route was specified to be carried: 'in a certain Mead called Mear Wood Great Meadow otherwise Clist Meadow, in the Parish of Burlescombe in the said County of Devon, belonging to William Ayshford Sandford Esquire, into and through the villages of Ashford and Sampford Peverell, to a certain Lane called Little Silver Lane,[10] in the parish of Tiverton aforesaid, whereby the Level might be preserved and carried on through the whole Length of the said Cut, and a great Expense and Inconvenience in Lockage and Consumption of Water would be avoided'.

The Act also laid down that as the line of the canal would pass inconveniently near to Sampford Peverell rectory the company should, within three years, either purchase or build some other suitable dwelling as parsonage house in lieu. It was further stipulated that the canal must not be built nearer than 100yds to the rectory house of Tidcombe in the parish of Tiverton without the permission of the patron of the living, the rector and the Bishop of Exeter. If, however, the permission was required, sought and given, the company would if necessary remove the parsonage house and build another suitable one on some other part of the glebe land.

As a safeguard against future financial delays and deadlocks due to non-payment of calls for money in respect of shares, the Act authorised the company to sue for sums unpaid, and provided that shareholders refusing or neglecting to answer the calls should forfeit 5s (25p) per share if not paid within 30 days of the appointed date, 10s (50p) if not paid within 60 days, and should forfeit the whole of the share or shares if the payment was not made within three months. Shares so forfeited were to be sold by the committee of management at public auction and the proceeds vested in the company for use in the canal's construction, or, in the absence of bidding at the sale and if the committee thought fit, sunk and merged in the rest of the shares of the undertaking.

DIFFICULTIES AND DOUBTS

By the middle of 1811 difficulties and doubts had arisen. On the bright side, Acts for the Bath & Bristol and the Bristol & Taunton Canals had been passed that year and Parliament had also established the privilege of the Monmouthshire Canal Company to export coal to Bridgwater free of duty. This improved trading prospects, for coal brought up the Tone would now be cheaper, and there would be less need for a canal extension to Uphill or Bristol. But achievement was becoming increasingly remote, due mainly to financial problems. The building of the eastern part of the main line, from Holcombe Rogus to Taunton, which, apart from some contracting for land, had been left in abeyance pending authorisation of the Bristol & Taunton, could not yet be considered, and even completion of the section under construction was in jeopardy.

Even so, the report which John Thomas, the superintendent, made to the committee of management on 22 June 1811, though not fully comprehensive due to John Rennie being ill, was still optimistic. It stated that the cutting at Holcombe Rogus and Canonsleigh 'answered the most sanguine expectations' and remarked that the committee, having attacked the greatest difficulty first, had by comparison nothing of importance in the remainder of the works, especially if the line were taken down the vale from Halberton to Cullompton, which besides being the easiest route possible would save a full 3 miles in length, making the whole canal from Taunton to Topsham, together with the branch, only 43 miles, with the branch to Halberton then being part of the main line. Thomas observed that in parishes at the east end of the canal 30,000 tons of lime were burnt annually and that carts came 30 miles by way of Tiverton to fetch it; with the improving state of agriculture twice as much was burnt as six years previously and additional kilns had been built but were not completely supplied. He visualised 60,000 tons of lime being carried annually at an increased rate of 3d a ton and pointed out that when the remainder of the line to Taunton had been made, coal could be carried on the canal to the kilns, and the whole vale to Topsham supplied with lime instead of it being brought from the south coast where duty-paid coal was used for burning it. He foresaw coal carried from Bridgwater to the interior of Devon, and corn, lime, slate and stone as back traffic.

Some members of the committee of management, Sir George Yonge, Peter Bluett, John Fortescue, John Kennaway, Thomas Fox and William Brown, in a report two days later, even more optimistically considered that Thomas might have included other prospective lime traffics, as well as the carriage of heavy articles. But other more pessimistic views prevailed, and elements of opposition existed within the company, as was apparent from a further paragraph of Thomas's report:

> 'I see no fairer speculation anywhere, to which the proprietors all agreed when the undertaking was commenced. The discouragement I have lately heard of must arise either from want of money, from an ill opinion of the management or from want of confidence in the speculation. From whatever cause it is found on enquiry to arise, I would recommend the Committee to meet the evil and I wish some of the members would go to the General Meeting in London for this purpose.'

Whatever idealistic future possibilities there might be, the immediate financial situation was becoming desperate. £90,000—more than a quarter of the sum authorised for the whole canal—had been spent by the summer of 1811 and there was much left to be done on the Tiverton branch alone. The high cost resulted mainly from the decision to lower the summit and then build the branch level with it, which meant more work in excavation and involved rock removal in making the deep cuttings. Another possible factor in increasing costs was the apparently high rate paid for the land purchased. No evidence has appeared to support any suggestion that the latter was due to owners of land holding out for large sums, and it was quite probably simply on account of the inflated land prices of the time—the later years of a long war—and the fact that the land required was of good agricultural value. Around £100 an acre was paid for land purchased, as the following examples of transactions show:

1813 Sale by Thomas Babb of Washfield, yeoman, to John Thomas and Henry Skinner of Grand Western Canal Company, lands in Halberton for £407 5s, viz:—

Part of Higher Close	3 acres and 25 perches
Strip, part of Lower Close	38 perches.

being parcels of Harwoods in Town.

1814 Sale by Charles Dally Pugh of Thorverton, to John Thomas and Henry Skinner, lands in Halberton, for £1,173 15s, viz:—

Part of Bottom Close	1 acre 1 rood 12 perches
Part of Copse Close	8 acres and 20 perches
Strip measuring	12 perches
Strip, part of Hill Close	3 roods and 8 perches
Part of Linhay Close	2 roods and 31 perches
Part of Barn Close	2 roods and 35 perches

being parcels of Bycott Farm, containing 140 acres which had been mortgaged by Charles Dally Pugh in 1812 for £3,000.

1815 Sale by Henry Laroche of Halberton, to John Thomas and Henry Skinner, lands in Halberton for £1,342 12s, viz:—

Davis's Close	1 acre
Strip measuring	2 roods and 32 perches
Part of Four Acres	31 perches
3 strips, part of Great Shutslade, measuring	3 roods and 10 perches, 3 roods and 4 perches and 30 perches.
Shutslade Plot	1 rood and 7 perches
Part of Nodbear	1 rood and 18 perches.
being parcels of Townsend and Jenkin's Tenement	
Four Acres of Bridge Ashland	4 acres 2 roods and 28 perches
Four Acres of Ashland	4 acres 2 roods and 23 perches
Part of Three Acres	2 acres 1 rood and 23 perches
being parcels of Aish Furland.[11]	

SUSPENSION

The General Assembly of shareholders—the forthcoming meeting in London to which John Thomas had referred—was held on 27 June. The occasion provided an opportunity for airing the current feelings of dissatisfaction and concern rife within the company—which the continuing war with Napoleon did nothing to lessen—and found the majority in favour of caution. The decision to apply the brakes, at least temporarily, was expressed in the resolution:

> That the small progress hitherto made in the Execution of the Works, with the increasing difficulty arising from the pressure and peculiarity of the Times, unforeseen at the Time of projecting the Undertaking, and the duration of which is beyond calculation, are valid and sufficient reasons for suspending the Undertaking, and that the Works shall for the present, be so far suspended, as may be consistent (in the Opinion of a Committee to be now appointed) with their final preservation.

Thus in June 1811 work on the canal came to a standstill. Deserted in mid-construction, the excavation reposed as an inconvenient eyesore, while most of the navvies no doubt found alternative employment under the engineer James Green, who was currently advertising for 100 canal cutters,[12] probably for work on the abortive Exeter & Crediton Navigation then in progress. Meanwhile the sub-committee appointed following the shareholders' meeting wasted no time in inspecting the canal, and at a meeting at Cullompton on 3 and 4 July, with Peter Bluett in the chair, formulated the following resolutions:

I That from the inspection of the Works the Committee are highly satisfied with the manner of Execution, as well as with the great progress that has been made.
II That certain parts of the Line are so nearly finished, and so circumstanced in other respects, as to make it impracticable to leave them in a state of preservation as to the Works, and with convenience to the Public, by any means so little expensive as the completing those parts.
III That it appears the other Works may be suspended and put in a state of security at an expense of about £9,000 to £10,000.
IV That in the event of such suspension, compensations must be immediately made to the Contractors and Undertakers to the amount of about £2,500.
V That the works may be slowly continued by collecting the Arrears of former Calls, and the last Call of £5 per Share (which may be made in 2 Instalments of £2 and £3 per Share) so to proceed gradually with the same until the Summer of 1812, thereby avoiding the Expenses of Compensation for Waste, and other necessary consequences of suspension; and at the same Time preserving the Works in a perfect and progressive State.
VI That it appears that the further Expense of completing the Line from Tiverton to Holcomb (its previous commencement) would be about £85,000.
VII That it appears the quantity of Lime to be annually carried on the Line to Tiverton, if completed, would amount, at the lowest calculation, to sixty thousand tons, and of Coal to about thirty Thousand, the Tonnage on which two Articles alone would amount to £9,600, at a Tonnage of three-pence per Ton per Mile. This Committee has purposely confined their calculation of Tonnage to Manure and Fuel, as the leading sources of Income.
VIII That the Rate of one-penny per Ton per Mile stated in

> the Act of Parliament on 1796 is very much below a fair Rate; —that three-pence per ton per mile would be a very moderate Rate, the price of land carriage, per Mile, by which the Lime and Coal are now conveyed, being nearly one Shilling and Six-pence, and that an Act for fixing a Rate of three-pence per Mile might, in all probability, be obtained without any difficulty; the Persons interested in the Carriage on the spot having expressed their full Approbation of a Tonnage so regulated.
>
> IX That these Resolutions, signed by the Chairman and other Members of the Sub-Committee, be transmitted as their Report to the Committee for the General Assembly of Proprietors adjourned till the 15th August.

The sub-committee report was incorporated in the report of the committee of management presented to the General Assembly at the City of London Tavern, Bishopsgate Street, London, on 15 August 1811. An appended statement by Samuel Woods, the committee of management's chairman, indicated that documents subsequently obtained made it apparent that the cost of immediate suspension of the works, which would involve fencing off the lands, levelling of spoil banks so as to return the lands to the owners, making good the public roads, erecting bridges, cutting back drains, and leaving the works in such a state that they might be resumed at the least possible loss, would amount to £27,000, which would require an additional call of £10 per share. (This in addition to the £20 per share already paid.) The cost of finishing the existing excavation and continuing the line to Tiverton was re-estimated at about £95,000. It was further expected that by completing the canal to Tiverton the revenue to be gained from lime, limestone and coal would in fact amount to over £10,000 pa, provided an Act for raising the tolls on the lines suggested could be obtained. The alternatives which the shareholders had to consider were: firstly, abandoning the concern, after spending nearly £90,000 and the produce of an additional call of £10 per share, with no benefits derived, or secondly, completing the canal to Tiverton, by which it would be productive of tolls to yield a probable 6 per cent on the whole sum expended, which would require, it was ascertained, calls of £30 per share. The second alternative, the chairman recommended, would enable the proprietors to obtain a reasonable benefit for the total sum of £50 per share invested.

The General Assembly gave the report a favourable reception and, though noting that the work under consideration involved only a small part of the whole canal project, was convinced 'that regarding the Time employed, the difficulty and importance of the Work in that Portion of the Canal which has first been put into Execution, the Excellence of the Work thus executed, the Economy in the Consumption of Land, and the quantum of Work done, a laudable Energy is evinced in the Committee of the last Year, and the Gentlemen who composed it are entitled to the renewed Thanks of the Proprietors.' Seeing the prospect of completion of the line to Tiverton as 'highly beneficial' to the proprietors, the meeting rescinded the previous resolution relating to suspension of the undertaking and further resolved that, in view of the increase in the value of land and of the price of materials and labour since the time of the original Act, an estimate of the probable expenses of completing the canal should be prepared prior to an application to Parliament for a further Act to authorise increased tolls.

Construction was therefore resumed, and at the next General Assembly in January 1812, progress towards gradual completion was reported, and promotion of the Bill for the increase of rates was in hand. There was backwardness, however, in the payment of calls; the treasurer had had to advance £1,500 and the solicitor was instructed to act against defaulters. The meeting was told:

> The committee are fully satisfied that if the canal be executed only between Holcomb and Tiverton the revenue thence arising will be adequate to afford a good interest to the subscribers and that whenever the canal shall be completed to Taunton a consequent increase of tonnage fully proportionate to the expenses may be confidently anticipated.

It would appear that by now the hopelessness of ever completing the canal to its originally-intended full extent was acknowledged, and the decision made to confine the project to the summit level under construction and its eventual extension to Taunton.

THE ACT OF 1812

On 20 March 1812 the company's third Act, 'An Act to alter and increase the Rates of Tonnage, authorized to be taken by the Company of Proprietors of the Grand Western Canal; and to amend the several Acts passed for making the said Canal' was passed. (52 Geo III c. 16.) By it commodities expected to be carried were grouped into three classes instead of the five listed in the previous Act, the rates for each being raised. They were as follows:

Coals, culm, coke, cinders, lime, limestone, ironstone, iron ore, lead ore, and all other ores, stones, tiles, slates, bricks, flag-stones, clay and sand and all articles to be used for manure and for repairing roads	3d per ton per mile
Rough timber, iron in pigs and bars, lead in pigs and sheets, tin in lumps and bars, charcoal, salt, corn, hay and straw	4d per ton per mile
Wrought metals, oils, wines, liquors, groceries, cheese, earthenware, and all other goods, wares, merchandise etc not specified	6d per ton per mile

(Cheese, it will be noted, had been moved, more logically, into the group which included groceries and other goods for human consumption.)

There were to be no changes in the wharfage allowances, nor in the charges for the carriage of persons, cattle and parcels, but the regulation prohibiting the passage through locks of vessels laden with hay, straw, corn in the straw, road repair materials and manure, except when water was flowing over the waste weir, was altered so as to exempt vessels of over 30 tons, or any of less tonnage which paid as for this weight. A succeeding clause clarified the company's liability for keeping bridges and their approaches in repair by defining the extent of the responsibility as being 'so far only as the wing-walls, arches and abutments thereof shall extend and no farther.'

THE STRUGGLE FOR COMPLETION

The immediate prospects looked a little brighter. Four months later, on 16 July 1812, the *Exeter Flying Post* advertised 'Canal shares for sale. Six shares in the Grand Western Canal, also six in Taunton to Bristol, now much improved in value by an act passed increasing the tonnage from 1d to 3d per ton. Apply Wescombe's Snuff Shop, Martin's Street, Exeter.' At the time, though, many shareholders were in default on their calls, and at the end of the year notice of the forfeiture of 205 shares, owned by 25 shareholders, was given.[13] The forfeiture was confirmed at a meeting of the committee of management held on 20 January 1813 and sale by auction of the shares ordered to take place on 5 February at the Auction Mart in the City of London. The shares were duly offered for sale but, as was reported at the committee meeting of 26 July, 'no person did or would bid for the same or any part thereof'. Consequently the shares were cancelled, and the names of their former holders expunged from the company's books.

At its meeting of 24 July 1813 the company appointed Joseph Champney to examine the whole of the accounts of expenditure from the revival of the undertaking (22 February 1810), to call for, examine and report on all vouchers, and to note and report amounts of salaries paid to past and current employees and the expenses which the company had incurred. Perhaps there had been suspicions of slackness and inefficiencies in the book-keeping; if so they were confirmed by Champney's report,[14] addressed from London on 23 September 1813. In a schedule annexed to the report Champney stated the expenditure annually, as ordered to be paid and as entered in the books of the company; but he found that, although the treasurer had received credit when the orders for payment had been made, in many instances they had not then been paid and some, amounting to about £220, were still outstanding. Also, the report stated, accounts were not opened in the ledger at the revival of the concern with the respective shareholders, or credit given as they paid their calls, nor were monthly balances made showing the state of the treasurer's account, or the defaulters. Champney recommended a proper and prompt system of debits and credits, emphasising 'that the Treasurer be credited the Monies paid by him, when the same are actually paid, and not when ordered by the Committee' and advised the making of monthly balances.

The schedule of expenditure, made up to 7 August 1813, showed the following totals for the 3½-year period:

	£	s	d
Contracts, materials, labour and surveying	87,779	4	8
Land	29,634	10	11
Shares in the River Tone	1,050	0	0
Damages to land etc.	2,301	14	7
Stationery, printing, advertising, stamps, postages and petty expenditure	458	15	6
Law charges	1,510	10	9
Salaries to persons having held or now holding employ under the company	3,376	8	10
Remunerations and presents	765	0	0
Rent, taxes, coals, furniture and sundries, for accountant's house and office	467	12	2
Charges for committees	410	5	3
Sundries not before enumerated	1,070	2	3
	128,824	4	11

A footnote to the sizeable sum shown for 'sundries not before enumerated' explained that, of the total figure given, £654 9s 6d appeared by the minutes and the London petty cash book to have been applied to the paying of deputations from

London to Devonshire, allowance to the London committees, salary to the clerk, engraving and printing of plans, etc. 'Shares in the River Tone' represent a part of the Tone Navigation's debt which the Grand Western Canal Company had acquired in order to gain an influence in the river which it was designed to join.

On 17 September 1813, in response to a company resolution of 24 June, John Rennie went to Devon to view the state of the works on the canal, having delayed this long so that he could present the latest possible picture of progress to the meeting being held on 25 September. For convenience, the length of canal from Lowdwells to Tiverton was divided into nine sections or 'lots' and in his report,[15] written from London and dated 24 September 1813, Rennie dealt separately with each as follows:

> The Holcomb lot has been very expensive, not only on account of the depth of cutting, but also, from a very great part being rock, in which many very productive springs of water have arisen; these it has been difficult to drain, from the great extent of level ground which lies to the North-west, in the pleasure Grounds of Major Brown, and the necessity of keeping the Canal full of water to boat earth from the deep cutting to the Vale of Cannons Leigh. It was therefore thought proper to make a Heading or Tunnel under the deep cutting towards Loudwell to carry the surplus water, which arises therein during the execution of the Works Eastward, into the River Tone, and this was nearly completed when I viewed this part of the Work on the 20th.
>
> The Canal is filled with water between the deep cutting above described and the Valley at Cannons Leigh, by which earth for this embankment, so far as done, has been boated, and except a small quantity from Haybeam that is required to complete the embankment, will be brought from the benchings and deep cutting yet to do near Whipcott. After the embankment is done, there will be about 27,000 cubic yards of earth and rock to remove, which has been agreed not to be done until it shall be determined to proceed with the Works towards Taunton.
>
> The masonry in this first lot is in general well executed—but

one bridge is still to be built—another to be coped— and an Aqueduct is to be made to convey the Whipcott Brook over the Canal.[10] The Expence of this, with the Earth Work yet to do, will amount to £6,057 5s. 3½d.

Lot second, (except part of the embankment at Cannons Leigh, and some other small matters) may be said to be finished; indeed there is water in the whole of it. The estimate to finish this lot amounts to £1,519 17s. 1d.

Lot third, or the Ashford lot is also in forward state, and there is water in several parts of it. The various small pieces of Work yet to do, are estimated at £1,692 2s. 0½d.

Lot fourth, or the Sampford lot, is not in so forward a state as the third lot, several works in this lot are still to do, the estimate of which amounts to £3,211 0s. 10d.

Lot fifth, or the Halberton lot, has been the most difficult and expensive piece of Work of its kind on the Canal. The Village of Halberton stands in the line where the Canal should have been made, to avoid which, the Works have been thrown into several pieces of high and expensive cutting; some being of hard rock, and others open porous sand. This has occasioned great delay by the whole requiring to be lined. Much cutting and lining is still to do, and several bridges to build, the expence of doing which is estimated at £13,419 18s. 11d.

Lot sixth, namely that West of Halberton, is also a difficult and expensive lot, and there are still many different Works to perform, the estimate for doing which amounts to £7,546 3s. 8½d.

Lot seventh, also is an expensive lot, principally on account of the open and porous nature of the soil, which requires this lot to be lined for its whole length, and puddle earth is difficult to be procured, the estimate for completing this lot amounts to £4,889 11s. 9d.

Lot eighth, is principally expensive from the open and porous nature of the soil; but here puddle earth can generally be got near to the place where it is to be used. There is a good deal of masonry still to execute on this lot, the estimate amounts to £4,710 1s. 1d.

Lot ninth and last requires also to be lined, but like lot eight, the puddle earth is near at hand, there is a good deal of masonry yet to do, the estimate of which amounts to £4,654 2s. 2½d.

The total estimate for completion of the works, including a

sum of £4,650 0s 1d included for contingencies, amounted to £52,350 3s which, Rennie feared, would make it appear as if little had been done in recent months. But this, as he pointed out, was not the case, as his estimates included certain articles previously omitted or unforeseen, in particular the great extent of lining required—which, due to irregular strata, could not be accurately ascertained until the ground was actually opened—and the deficiency of puddle earth for the job. However, Rennie was in great hopes that what was now estimated would suffice and he concluded with the paragraph:

> The Works now appear to me to be in a fair train for completion, and I think with due exertion, the Canal may be opened as far as is now under execution, namely, eleven miles, by the month of September next, and gives me much pleasure to state that the Works, so far as executed, are in general very well done.

Construction work continued through the remainder of 1813 and into the summer of 1814. On 7 May 1814 the company received counsel's opinion on a problem which had arisen from demands by parishes on the line of the canal for payment of land tax and poor rates in respect of land being dug. The company doubted its liability and sought counsel's advice, which was that the company was not liable for relief of the poor until the canal was completed and had begun to yield a profit, and regarding land tax, that the proper procedure was to make a fresh assessment each year according to the value in that year.[17] May 1814 also saw the production of a report by the engineers W. Wallace and John Easton on suggested lines of horse railway from Tiverton to Chulmleigh via Cadeleigh and Bickleigh, or via Crediton, to connect with the canal. The area which would be served currently obtained its supplies of lime—probably 30,000 tons a year—from Barnstaple, Topsham and Canonsleigh, and it was thought that the project might cause all supplies to be got from Canonsleigh, transported via canal and railway; but there were engineering difficulties and not much enthusiasm, and the idea did not materialise.[18]

Page 53 (*above*) The Lowdwells terminus and remains of the lock which connected the two lengths of canal; (*below*) Silk Mill bridge, which carries the Bishop's Hull–Staplegrove road over the line of the former canal west of Taunton.

Page 54 (*above*) A section of the abandoned canal south of Norton Fitzwarren, still containing water; (*below*) a section of the abandoned canal immediately above the Allerford lift, with the renovated lift-keeper's cottage on the right.

On 16 June 1814 members of the committee travelled on a section of the canal, passing in barges over the embankment at Canonsleigh. On 29 June the committee reported to the General Assembly that there had been delay owing to the unprecedented severity of the winter, but that the canal should be navigable by the end of August. Two sections at the Tiverton end were still unfinished, with some cutting and puddling yet to be done. Wharves had been constructed at Sampford Peverell and Whipcott at the expense of the lime burners, another at Sampford and one at Halberton at the company's expense; the Tiverton wharves were in preparation and were to be paid for by those who were to use them. The report stated: 'It was thought expedient to provide a few barges for the immediate occupation of the trade that no delay might arise in its commencement; these will probably be purchased by the parties who wish to use them'. Fewer defaulters were reported but it was stated that estimates, even as corrected by Rennie, had been exceeded, and a call of £6 for completion was made.

On 25 August 1814 the first barge to travel the length of the canal arrived at Tiverton laden with coal, which was consequently reduced in price by 3d a bushel.[19] What was thus seen as an advantage to both the rich and poor of Tiverton had, however, been at great cost to the canal company. The expense of constructing this, the first section, had necessitated the raising of £244,505 in a succession of calls totalling £79 per cent on 3,095 shares.[20] (It appears that, of the nominal capital of 3,300 shares of £100 each—2,200 initially, augmented by a further 1,100 created by a deed of covenant of 1810—no more than 3,095* were actually created and issued.) Total cost so far, therefore, has already far exceeded the estimate of the 1796 Act for completion of the whole canal.

* The records vary between 3,095 and 3,096.

CHAPTER 3

Schemes for Extension

THOUGH navigable by the late summer of 1814, the canal undoubtedly still needed finishing work, and probably a few months elapsed before regular transport on it was in full swing. On 22 December that year a press advertisement[1] under the names of Samuel Woods, Robert Sutton, Samuel Drewe, Cornelius Buller and L. R. Mackintosh gave notice of a meeting called for 12 January 1815 to make bye-laws. Following the customary pattern, these, which were to come into force from 1 March, included the prohibition of the use for hauling of horses 'or other beasts' unmuzzled, also use of the navigation without licence before 4 am and after 9 pm from 1 March to 1 November and before 6 am and after 6 pm in the remainder of the year, and forbade bathing in the water and the depositing of dead dogs, cats or other carcases.[2] Tiverton wharf was in commercial use by the beginning of 1815, with the sale of lime from it advertised by Dunsford & Browne in the *Exeter Flying Post* of 12 January, and by Hugh Talbot in that of 23 February.

Once in use, the waterway brought about a great and rapid extension of trade from the quarries at Holcombe Rogus and Burlescombe, particularly that of Canonsleigh quarry close to the village of Westleigh, but to nothing like the extent of the promoters' expectations, nor to any degree approaching the annual figure of 60,000 tons of lime alone estimated by the subcommittee in 1811. The entire traffic was in lime and roadstone travelling westwards—i.e. from Holcombe Rogus to

Tiverton, with none in the opposite direction, and the only immediate beneficiaries from the opening of the canal were the quarry owners and the agriculturalists who took advantage of the new facilities to use lime more liberally and over a wider area. The canal company tried to create a trade in coal and experimented with coal brought overland from Taunton to Lowdwells and carried from there to Tiverton on the canal, but this proved no cheaper than land carriage straight through, and hopes for a coal trade were reluctantly relinquished and not resumed until the later extension of the canal to Taunton. In the twenty years before that could be achieved never in any year did the gross tonnage amount to £1,000—reaching less than a tenth of the 1811 prediction—the average return per annum in tolls being only about £600. Needless to say, no dividend was paid.[3]

BEVAN'S PLAN

In the years following its opening several projects were put forward for continuing the canal to Taunton. One such scheme was the object of a report by Benjamin Bevan, presented on 6 October 1818 and considered at a meeting of the committee on 31 March 1819. Bevan's plan envisaged 30 small locks to effect the rise in level from Taunton to Lowdwells with the canal itself being constructed 13ft wide at the bottom, 26ft at top water and 30ft at top bank, with depth 4ft. So as to accommodate singly boats which were capable of passing the locks on the River Tone in pairs, he suggested that the locks should each be 52ft long and 6ft 9in wide. His estimate, placing the locks in pairs wherever practicable, is shown at the top of the following page.

The committee optimistically modified and contracted individual items of the scheme and by so doing reduced the total estimate to a figure of £70,000. This it considered against an estimate of revenue calculated 'upon the most accurate information', drawn up by the 'Sub-committee in the Country'

	£
Purchase of land and damages	10,660
Cutting canal, level, deeps and embankments	25,547
Locks, 30	24,000
Bridges, 2 at £450, 15 at £350, and 20 at £300	12,150
Aqueducts, culverts and trunks	3,747
Towing paths, fencing and gravelling	2,825
Lock-keepers' houses	1,750
Stop-gates, waste weirs, watering-places etc.	550
	81,229
Contingencies	13,771
	£95,000

on the basis of Bevan's report. The relevant extract from the country committee's minutes read:

It appears that about 32,000 Tons of Coal and Culm are brought to Taunton on the River Tone, and that there is no doubt, that including the Culm carried to the Lime Kilns, 20,000 Tons (nearly in equal quantities each) are carried Westward of Taunton; we have no doubt that the increase of Culm used in burning Lime will double the present consumption, and as Coals will be so much cheaper on the West end of the Canal by Water-carriage, the Sale will be extended much further Westward, take this at one half more, the whole quantity will be 35,000 Tons. The distance to the Lime-kilns is 13 Miles, the coal will be carried from 7–22 Miles, taking the average at 13 Miles, at 3d per Ton per Mile, the amount will be

£5,687 10s 0d

The Trade in Lime produced from the end of February 1818, to the end of February, 1819, £700 0s. 11d, this is at two thirds of Toll, but when Culm is rendered cheaper by Water Carriage, it will bear the full Tonnage, which is

£1,050 0 0

The expectation is that this Trade will greatly increase, but call it double

£1,050 0 0

Lime will be carried Eastward as well as Westward, and Coals and Culm will be carried by the Canal to a Country now supplied from Watchet, suppose

£1,200 0 0

There are Slate Quarries at Wivelscombe, 7 Miles from Harpford Bridge, which it is supposed will produce Tonnage from carriage both ways on the Canal; there is also a probability that great quantities of Road Stones will be carried from various parts of the Canal, for these and all other Articles, such as Corn, Timber and Merchandise not calculated on, say

£1,000 0 0

£9,987 0 0

The available resources of the company at this time appeared to be 3,096 shares at £21 per share still uncalled for, totalling £65,016, land unsold valued at £2,000, any benefit resulting from the sale of lapsed shares, plus any progressive revenue arising from the canal's execution.

The plan went forward for consideration by the General Assembly on the following 24 June, but was evidently felt to be financially impracticable. Bevan's fee of 50 guineas was paid and the project subsequently dropped.

INCEPTION OF THE BRIDGWATER & TAUNTON CANAL, AND THE PROJECTED SHIP CANAL

In 1822 there was a revival of interest in the Bristol & Taunton Canal scheme, for which an Act had been obtained in 1811 but on which no start had been made. In about 1818, when there were no immediate prospects for the continuation of the Grand Western Canal to Taunton, the Bristol & Taunton Company, which saw advantage in obtaining a stake in the opposing River Tone, bought from the Grand Western Company shares in the Tone debt which the Grand Western had itself acquired a few years previously. By 1822, though powers to construct the middle section—from Clevedon to the River Parrett—had lapsed, the Bristol & Taunton shareholders felt that the time was ripe for commencement of the western section, from Taunton to the Parrett at Huntworth, which was still authorised. It was felt that this, potentially viable on its own, would provide an improved alternative to the River Tone; moreover, it would join a tub boat canal then being promoted from it to Beer on the south coast. In September 1822 work on this western section of the intended Bristol & Taunton Canal commenced, but continued for only a few months, due to legal difficulties which halted proceedings. When, in spite of petitioning by the Conservators of the River Tone who feared competition, a fresh Act was obtained in 1824, the old line east of the Parrett was abandoned, and the name of the company was changed to the Bridgwater & Taunton. At this point the Bridgwater & Taunton Company approached the Grand Western on the possibility of the latter extending its route to Taunton, or alternatively of allowing the Bridgwater & Taunton Company itself to construct the linking section, but no decisions were taken due to a new pre-occupation, that of a plan for a proposed ship canal between the English and Bristol Channels that had replaced the tub boat scheme of 1822.

The plan 'for making a Ship Canal between the Bristol and the British Channels', latest in a succession of inter-channel schemes, was considered at a meeting held in London on 9 June 1824 and surveyed by Thomas Telford, assisted by Captain, later Sir George, Nicholls and James Green, who had earlier surveyed the tub boat canal. The engineers' report, presented and adopted at a meeting in the following December, recommended a canal 15ft deep with 30 locks, suitable for carrying 200-ton vessels from Stolford on the north Somerset coast through Creech St Michael, Ilminster and Chard to Beer on the coast of east Devon. The cost was estimated at £1,712,844, the expenses at £22,000 pa, and annual revenue at £210,847. Heated opposition resulted and doubts were expressed as to the practicability and viability of the scheme. It had its supporters, however, and the Bridgwater & Taunton Canal Company, seeking to take advantage, agreed early in 1825 on a clause for inclusion in the English & Bristol Channels

Ship Canal Bill, by which its own unfinished canal, together with land and construction equipment and the Tone debt, would be bought by the Ship Canal Company, the purchase price of £90,000 being agreed pending the Bill's ratification. The Grand Western Canal Company took exception to this arrangement and with others prepared a case against the Bill for introduction on its third reading.

The Grand Western company's case was that it had itself been empowered to make a canal from Topsham through Tiverton to Taunton, with a view to opening a communication between the two channels, and it was pointed out that in 1811 a company was formed and Acts of Parliament obtained for making a canal from Bridgwater to Taunton to join the Grand Western and effect such a communication. This canal, it was stated, was nearly completed, and its company was bound, by an Act passed in 1824, to complete it within 3 years. Four main points were submitted on the part of the Grand Western proprietors. The first drew attention to the fact that on payment of the £90,000 the liability of the Bridgwater & Taunton Canal Company to complete its canal within the required period, or at all, would cease, and that if the Ship Canal project should subsequently fail, the Grand Western would lose all the sources of profit held out to it by the Bridgwater & Taunton, consideration of which had been the principal inducement to the Grand Western Canal's commencement and execution, on which, it was stated, upwards of £200,000 had already been spent. It was contended that the Ship Canal Company ought to remain bound to complete the communication between Bridgwater and Taunton before such liabilities ceased.

The second point made was that the Bridgwater & Taunton Company's Bill provided for a toll no higher than 2d per ton, on which basis the Grand Western's scale of expenditure and profit had been calculated, and that the Ship Canal Company ought to be bound to take no higher toll on barges and boats navigating any current or future communication between Bridgwater and Taunton and along any part of the Grand

Western. Thirdly, the tolls which the Ship Canal Company was proposing were so high that the Grand Western conceived it would be unable to use it, and thus would be prevented from completing its undertaking while losing the large capital already expended. Lastly, it was stated that the excessive tolls being proposed by the Bill represented a great addition to those previously suggested and circulated in the Ship Canal Company's printed reports and that the Grand Western Company and others had been prevented from seeing the extent to which their interests were affected. Two clauses were put forward by the petitioners for inclusion in the Bill, the first that the Ship Canal Company should be compelled to complete the navigable communication between Bridgwater and Taunton within the three years, and secondly that there should be no alteration in the rates of toll already granted in the Acts for the Bridgwater & Taunton Canal.

Amid much enthusiasm the Act for the English & Bristol Channels Ship Canal (6 Geo IV c. 199) was passed in July 1825. Most of the necessary money had already been subscribed, but, with a change in the financial climate, the prospects dimmed, and after an initial start in the surveying, work stopped. In 1828 the company considered a modification of Rennie's old scheme of 1811; after that all activity ended.

In the meantime the Bridgwater & Taunton Canal had been completed and opened on 3 January 1827. In August that year a fresh suggestion, for a small ship canal, came from Charles Dean, a surveyor and engineer of Exeter. His plan was for a cut from the mouth of the River Parrett to Bridgwater, where there would be a dock, then widening and deepening of the Bridgwater & Taunton Canal and construction of a new canal from Taunton to Lowdwells, with an extension from the existing Grand Western by Cullompton, Bradninch and Stoke Canon, crossing the Exe to join the Exeter Canal at a basin then being made. There would also be a branch to Crediton. Dean envisaged a canal suitable for carrying vessels of 80–100 tons, and estimated the relative cost of construction at

£500,000 or £700,000 respectively.[4] But the scheme failed to stir sufficient interest, and was shelved.

MATTERS OF FINANCE

Although, as has already been stated, the amounts received in tolls after the canal's opening fell far short of earlier expectations and produced no profits to shareholders, the company's working accounts showed a modest but gradually increasing credit balance as the 1820s advanced, reaching over £1,300 in 1829. While the known annual tonnage receipts varied between £381 and £753, other income came from such sources as the sale of tramplates—presumably from the dismantling of construction works—and from rents, which would have included £63 annually from Hugh Talbot, yeoman, of Holcombe Rogus, and a similar sum from Henry Dunsford, a banker of Tiverton, in respect of wharf space on the north side of the canal at Tiverton for which in 1820 each was granted a 21-year lease by the company.[5] On the debit side, repair and preservation of the canal and works stood at a steady annual figure of £275, and represented the biggest expenditure in any one year except for a total of £422 for law expenses in the year ended June 1824, no doubt incurred in opposing the Ship Canal Bill. In the year ended June 1829 £105 was spent on building limekilns and walling at Tiverton and Ebear, the former evidently being the subject of further leases negotiated in 1830, one with John Potter, a lime burner of Burlescombe, who was let a limekiln and lime wharf on the north side of the canal at Tiverton at an annual rent of £22 12s 6d (£22.62½p), and the other with John Talbot of Holcombe Rogus, also a lime burner, for similar premises over the same period at £23 per annum.[6] A statement of the company's capital assets and liabilities at 23 June 1829 showed effects valued at a total of £4,618, including invested stock worth £1,050 and land at the 1815 valuation figure of £1,760. Debts or claims included £1,460 as the 'probable expense of building Sampford Peverell parsonage', a requirement of the 1811 Act which as yet had not been fulfilled. The probable balance was calculated at £2,570.

ENTER JAMES GREEN

It was in 1829, when failure of the lime trade to achieve the anticipated volume had been acknowledged, and any chance of future success was seen increasingly and inevitably to lie in the extension of the route to Taunton and connection there with the Bridgwater & Taunton Canal, that the engineer James Green made his entry upon the Grand Western scene. Green, at 48 years of age, had already shown himself a man of considerable talent. Born in Birmingham, he had acquired his early engineering knowledge from his father under whom he worked until he was 20, following which he was employed under Rennie on assignments in different parts of the country. 1808 brought him to Devon, when, after submitting successful plans for rebuilding Fenny Bridges in the east of the county (the previously reconstructed bridges having collapsed after only 18 months' use) he was appointed as Devon's Bridge Surveyor. Ten years later he was elevated to the post of Surveyor of Bridges and Buildings for the county of Devon, which he held until 1841. In addition to his official duties Green undertook much private work during his years of appointment in Devon, as was customary for county surveyors at that time.[7] He was engineer to the projected Exeter & Crediton Canal—started in 1810 but shortly afterwards abandoned—and from 1820–7 was engaged in enlarging and extending the Exeter Canal.

But it was his grasp of the challenge of canal building in the south-west's hilly countryside, his advocacy of small tub boats and his methods of lifting which made James Green one of the notabilities of the age in westcountry engineering. In 1818 he presented a report to the subscribers of the Bude Canal,[8] and from 1819–25 supervised that canal's construction, most of it to take 5-ton tub boats. The line included six water-powered inclined planes. The methods by which these planes operated were not Green's own ideas, owing much to the earlier Robert Fulton,[9] but they were ingeniously contrived; five of the planes were powered by waterwheels sited at the top and the use of endless chains, while the sixth, that at Hobbacott Down, which rose an impressive 225ft in a length of 935ft, was unique. At Hobbacott were two wells sunk at the plane's summit, equal in depth to the vertical rise, in each of which a huge bucket, suspended on a chain wound over a drum and containing 15 tons of water, provided the motive power as it descended the depths of the well, the water being discharged into a drainage adit as the bucket touched the bottom. Though a 16hp steam engine was installed for use during breakdowns (which were frequent) the Hobbacott bucket-in-the-well system worked effectively when all was in order, and was extremely swift, raising loaded boats the length of the plane in 4 minutes—half the time the steam engine required. While the Bude Canal was under construction Green had, as we have seen, in 1821 surveyed a line for the proposed inter-channel canal across Somerset and east Devon, recommending the use of tub boats and inclined planes (a scheme which did not materialise but which was a fore-runner of the later Chard Canal), in 1823 produced a plan for the Liskeard & Looe Union Canal suggesting the same means, and during 1824–5 supervised the building of Lord Rolle's canal from Bideford to Torrington which carried tub boats and had a water-powered incline. Green's career was to culminate—alas erringly—with his work on the Grand Western Canal's second era of construction.

GREEN'S FIRST REPORT (1829)

Early in 1829 James Green, seeing the prospects for applying his methods and experience as a solution to the problem of the Grand Western, prevailed on a number of the canal's subscribers in the Exeter vicinity, where he was well known, to call a meeting to consider a plan he was proposing for providing a communication between Taunton and Holcombe Rogus. The meeting was held at Exeter on 1 May that year.

In this, his first report to the Grand Western Company, dated 28 April 1829,[10] Green observed that the existing length of canal was 6ft 6in deep, 24ft wide at the bottom, 46ft wide at the top, had 16 arched and 5 swivel bridges and a clear waterway of 18ft beneath them. But the width of locks on the Tone Navigation was only 13ft 9in, and the width of the Bridgwater & Taunton Canal, which it was probably expected would have been the same as the Grand Western, was in fact only 14ft under the bridges, so that the width of a new length of canal need not be greater than this. Remarking that the carriage of limestone, culm and coal had been disappointing because of the great reduction in the cost of land carriage, largely accounted for by the change from war to peace, Green wrote:

> 'it is now found that the rate of land carriage of coal and culm from Taunton to Tiverton and the adjacent districts is so low that scarcely any of these articles are put into Boats at the North East end of the Canal and borne thence to Tiverton, whilst the cost of carrying culm and coal to Tiverton even at the present reduced rate of land carriage is such as to prevent any considerable quantity of limestone being taken from Holcombe Rogus to Tiverton to be burned into lime, the state of agriculture not allowing of a remunerating profit to the lime merchant. The expectation therefore of receiving any income on this part of the Canal from the sources anticipated must remain unanswered until Coal, Culm and other articles can be conveyed from Taunton to Tiverton at a price much below the present rate of land carriage and this can only be accomplished by forming some mode of communication between the Canal near Loudwell Mill and the Bridgwater & Taunton Canal near Taunton.

Having made many enquiries, Green reported upon two alternatives, a railway or a canal. Considering the railway, he stated that if this was to rise by a gradual incline the route would have to be very roundabout with expensive cuttings and embankments, otherwise there would be steep ascents which would need stationary steam engines which, in addition to the cost of construction, would be expensive to maintain so far from coalfields. There would also be the inconvenience of the double transhipment of cargoes.

Concerning the canal alternative, Green listed three points for attention; the expense of making, which would depend on the size; compensation for land, which, in view of the long time since the original Act, would involve higher payments having to be paid for improvements, and the obtaining of water. He wrote:

> It may be necessary to alter the present Parliamentary line, for in some cases the Locks are so placed as to be almost impracticable, in some parts the line may be varied with advantage, within the distance allowed by the Act, but in one instance a serious alteration should be made, for to render the Canal perfect it should communicate with the Bridgwater and Taunton Canal below the town of Taunton, instead of the River Tone above that town, probably this might be effected by private arrangement.

Green referred to the estimated dimensions earlier suggested by Bevan (in 1818) and the then adjusted cost of £70,000, pointing out three objections: firstly, the great waste of water with so many locks and a predominantly upward trade; secondly, that the width at the bottom was too small to allow passing of vessels without damaging the side of the canal; and thirdly, that barges would probably be too big to pass through the Parrett. Cost of land had previously been estimated at £8,360 but would now be more. And consumption of water, at 8 tons to 1 ton of cargo, would provide for a maximum trade of only 45,924 tons without reservoirs, water having to be brought from Lowdwells Mill. The revised cost of this particular scheme would thus probably be £100,000, with the transit of only 20 tons of cargo per horse.

Green's suggested plan was for a canal with inclined planes. It would start at the Bridgwater & Taunton Canal, though boats could use the Tone, and would avoid the Nynehead and other expensive properties, greater latitude for the line being possible. He recommended 13ft width at the bottom, 23ft at water surface and 3ft depth, with two, or at the most three, inclined planes worked by water, 3 tons of which would be required per ton of cargo. Boats 20ft by 6ft, carrying 5 tons and drawing 2ft of water—almost exactly similar to those in use on the Bude Canal—were advocated; six of these would be drawn by one horse on the levels of the canal and each would take 5 minutes to pass over an inclined plane. They could navigate the Parrett in pairs and the Bridgwater & Taunton Canal in sets of six at once. Consignments of timber could be floated on the water and pass up the inclines on special carriages. Green noted that canals had existed on this principle for 30 years in Shropshire, worked by steam where coal was cheap (a reference to the Shropshire and the Shrewsbury Canals) and stated that he had himself built the Bude Canal of 39 miles with inclined planes worked by water, and also the Torrington, in the same way; 'they have been found to answer perfectly well' he wrote, noting that the Bude Canal had an ascent of 400ft and a descent of 180ft.

Such a plan, Green pointed out, would have the advantages of cheapness, of using less land and allowing variation of line, and would use less water and be quicker through having inclined planes. In dry seasons 3½ft of water could be taken from the upper level of canal to work the planes, which would be replaced by rain and the Lowdwells springs and be sufficient, it was estimated, for allowing passage of 164,962 tons of cargo a year. However, small steam engines could be used as a last resort if there was too much traffic for water. The cost he estimated at £50,000 or probably less 'on a minute and careful survey', and Green wrote: 'I will if required engage to find responsible persons who will undertake the work with adequate security for that sum'. Regarding the 1819 estimate of returns of £9,987 10s per annum, Green observed that this had been made on the supposition that the annual carriage on the

River Tone to Taunton of coal and culm was 32,000 tons, and he noted: 'It is now found that the tonnage on the Bridgwater & Taunton Canal and the River Tone amounted in the last year to near 60,000 tons. It is therefore probable that the returns on the Grand Western Canal would much exceed the calculation made in 1819'.

In remarks which he made on the original plan for a canal from Topsham to Taunton, Green stated that in order to be of sufficient dimensions to carry seagoing craft the cost would have been prohibitive, and if smaller the cost of transhipment would not have competed with vessels sailing around Land's End. In any case the entrance into the tideway of the Exe could not have answered the expected purpose. But he suggested that after the plan he was presenting had been completed, there should be a continuation to Exeter on the same principle at an estimated cost of £70,000, taking the line from the Tiverton branch to Cullompton, Broadclyst and north-east of Pinhoe to join the Exeter Canal (recently improved to take 300-ton vessels) immediately below Exeter. As trade would be chiefly from the north little water would be required for the inclined planes.

The subscribers' meeting of 1 May adopted Green's report, and at the following annual shareholders' meeting held on 25 June, the following resolution was passed: 'That it is expedient to make a canal communication (as proposed) providing the same can be made and in every respect completed including the purchase of land and damages attending the execution of the work at any expense not exceeding £50,000.' James Green was asked to make a general survey of the parliamentary line and to confine himself to it, though turning his attention to any such deviations from it as would materially facilitate the execution or lessen the expense.[11]

GREEN'S SECOND REPORT (1830)

Green's subsequent and second report was dated 2 March

1830 and showed some changes from his earlier plan of the previous year. One was that the originally authorised line from the south bank of the River Tone at Taunton—which necessitated crossing the Taunton mill leat and the River Tone—was, for the time at least, retained, while in addition Green had altered his ideas for lifting, suggesting now that there should be just one inclined plane and in addition seven perpendicular lifts, with boats of 8 tons being carried. The cost of actual construction he estimated at £49,324, while added amounts for contingencies and payments for land and damage brought the total figure to £61,324. The abstract of the report read as follows:

> The line which I have chosen for the purpose of effecting a Canal communication between the River Tone, at Taunton, and the Summit Level of the Grand Western Canal, near Holcombe-Rogus, will be found by the plans to be a remarkably direct one. It commences at the River Tone, about a quarter of a mile above the Bridge at Taunton, and follows nearly the Parliamentary Line throughout.
>
> The Line is more advantageous to the Landed Proprietors than any hitherto laid down, in several instances it forms a convenient division of property, and by which many bridges (at all times impediments to a Canal) are avoided, as well as heavy expenses in severance damages. It is also more out of the reach of floods, and will be less affected by them than if the original line were prosecuted.
>
> I have found it necessary, in consequence of the very gradual rise of the land in the first five miles of the Canal from Taunton upward, to make in that distance four perpendicular lifts. I have also made some modifications of the respective heights of the several lifts, which have been so regulated as to suit the features of the Ground and diminish expense, whilst the whole number of Lifts will remain the same, viz Eight, over which a Boat with eight Tons of cargo may be conveyed in twenty-four minutes; and supposing a Horse will generally draw on the Canal four Boats with thirty-two Tons, the time required for the whole ascent of thirty-two Tons of cargo, will be only about one hour and a half. The position and rise of the Lifts laid down in the longitudinal Section, and as marked on the plan, will be as follows:

Page 71 (*above*) The aqueduct near Nynehead which carried the canal over the River Tone; (*below*) the iron trough of the aqueduct by which the canal was carried over the River Tone.

Page 72 (*above*) Remains of the Nynehead lift which raised the boats a height of 24ft; (*below*) close-up of the Nynehead lift masonry, showing holes into which the metal framework was bolted.

1.	Near the commencement of the Canal at Taunton, in order to gain sufficient height to pass the Canal over the Taunton Mill Leat without interference therewith, and also to pass the Canal by an Iron Aqueduct over the River Tone, near Bishop's Hull	20 feet rise
2.	Lift near Norton, or Allerford Brook	16 ditto
3.	Ditto near Hillfarrence Brook	19 ditto
4.	Ditto near Trefusis Farm, which is of a sufficient height to pass the Canal a Second Time over the River Tone, by an Iron Aqueduct near Ninehead	38 ditto
5.	Lift on the North* Side of the Tone near Ninehead Court Lodge	24 ditto
6.	Lift near Winsbeer Quarry	18 ditto
7.	An Inclined Plane near Wellsford	81 ditto
8.	A Perpendicular Lift near Greenham, which attains the Summit Level of the Finished Part of the Canal	46 ditto
	Total Rise	262 feet

The Country through which this Canal will pass is favourable to the execution of such a work, the Ground being generally of easy cutting, and although much Lining and Puddling will be required, the Line of the Canal will furnish abundance of Earth fit for that purpose; good Stone for building may be procured within a moderate distance of the Canal, and in many convenient places good Earth will be procured in, or adjoining the Line for making Bricks. I have, however, designed the Aqueducts Public Road Bridges and Viaducts for carrying Roads under the Canal, (which may be done in many instances, and prove a great convenience to the Public as well as to the Canal) to be principally of Cast Iron; the cheapness of this Article, its convenience in fixing, and its durability, rendering it for these purposes most desirable.

I have also ascertained that the works may be commenced at Taunton, and prosecuted upward, whereby the Canal may be made productive in many parts before the whole is finished, particularly at Wellington and some intermediate places.

* 'South' is meant.

5

I have estimated the cost of executing the Work in eight separate sections, which may be contracted for in so many distinct Lots, or by two or more taken together, the total amount of which is £49,324 6s. 7d.

This estimate has been made with every possible attention and detail, and I have no doubt it will be found adequate to the purpose. The work has been valued at liberal prices, and as the cost of labour and every sort of material is now very moderate, I have good reason to think it may be executed for less than that sum; but as some trifling expenses may occur, which it has been impossible to foresee, it may be prudent to add to the above sum, £2,000. The cost of land and damages cannot exceed £10,000, making a total of £61,324: 6s. 7d.

To the Chairman and Committee of Management of the Grand Western Canal Company. JAMES GREEN

On 15 March Frederick Leigh, the Company's principal clerk, issued from Cullompton a notice calling a special assembly of proprietors to be held on 13 April. The main item of business was to consider Green's report and to decide whether or not to proceed with the works to complete a junction between the summit level and the River Tone according to the plan and estimates proposed. It was also to consider rescinding a resolution passed on 29 June 1815 which restricted the company from making any additional calls and to authorise the committee of management to adopt various measures as might be necessary. The plans, models and sections of the proposed waterway were on view for examination by the proprietors at the George & Vulture Tavern, near Cornhill in the City of London, on the morning of the day of the general meeting, when the chair was taken by Samuel Drewe.

Approval was given by the meeting to Green's report and the committee was authorised and instructed to put into effect the 'perfect execution' of the work, with the greatest economy and keeping the sum within £65,000.[12] The other items on the agenda also were given the necessary sanction—it was noted that a secretary would be required, also a London office—and, by way of reassurance, the assembly was told:

'By what means effectual provision can be made against any excess of expenditure beyond the estimate now laid before the General Assembly must remain with your Committee an object of increasing solicitude. They are not aware of any other means of guarding against such excesses, but great care in the formation of the original contract and constant watchfulness against the creation of any collateral demands. Instructed by former experience the Committee are disposed to retain in their own hands not merely the superintendence, but the detail of every contract and expence, the examination of accounts and the payment of monies, instead of entrusting as before the entire management to a sub-committee in the country.'[13]

CHAPTER 4

The Second Stage of Construction

THE 'REVIVAL OF THE UNDERTAKING'

FOLLOWING the unanimous agreement on James Green's report, the general feeling was of optimism. The committee considered the scheme should produce a 10 per cent return on capital and some proprietors even thought 15 per cent.[1] On 14 April, the day following the special general meeting, the committee made a call for £5 per share to be paid to J. W. Lubbock & Co before 10 June, and this was followed later by two further calls of similar amount, and two of £3 each, the last being payable on 23 August 1833. But before the end of 1830 plans for the commencement of the work were going forward and tenders were being invited. The draft for one such notice read as follows:

> Grand Western Canal
> To Canal Contractors, Masons and Builders.
>
> Such Persons as may be desirous of contracting to execute the cuttings, Embankments, and Buildings in Lots 7 and 8 of this Work, extending from the present Canal, near Holcombe Rogus, towards Taunton, about two Miles, two furlongs, and four Chains, may see the Plans and Specifications of the Work to be performed on application at the Office of the Company at Landcocks, near Wellington or at the office of Mr Green, the Engineer, at Exeter. Tenders properly sealed and Indorsed, 'Tenders for Works on the Grand Western Canal' may be sent

to Mr Green on or before Wednesday the 5th January, after which Day no Tender will be received. The Committee will decide on the Tenders at Landcocks on Wednesday the 12th January at twelve OClock at Noon. Adequate security will be required for the due performance of the Contracts; and the Committee do not pledge themselves to accept the lowest Tender.

Dated December 13th 1830 Signed
Fred: Leigh principal Clerk

Either towards the end of 1830, or early in 1831, at a time when there was hostility between the Bridgwater & Taunton Canal Company and the Conservators of the River Tone, the Grand Western Canal Company decided to make a deviation in its route at Taunton from the parliamentary line laid down in the Act of 1796. The ensuing dispute involved the Bridgwater & Taunton Company forcing a connection from its termination at Firepool into the River Tone close by, so that it could send barges on up the river to a wharf at Taunton bridge, and retaliation by the Tone Conservators, who kept the canal short of water and did all they could to keep the canal company out. During 1830 litigation was in progress and no doubt the Grand Western Company felt its position would be more secure if connection could be made directly with the Bridgwater & Taunton Canal instead of with the river. This was therefore made possible by agreements and purchases of land from local landowners and on 16 June 1831 the *Exeter and Plymouth Gazette* reported that work had started on the extension of the canal both between Holcombe Rogus and Wellington and near the junction with the Bridgwater & Taunton Canal at Taunton, where there was to be a 'handsome, lofty aqueduct over Rowbarton Road near Mr Liddon's'. Agreement reached between the two disputing bodies at Taunton resulted in an Act passed in 1832, one of the clauses of which required the Bridgwater & Taunton Company to build a direct canal between the Tone and the Grand Western. This was cut in 1834 from a point just below French Weir north-westwards to Frieze Hill, though there appears no evidence that it was much used, if at all. There was, it seems, inevitably a lock connection between the river at French Weir and the cut, which was quite narrow and ran generally parallel with the river and at certain points very close to it, but an extremely sharp right-angled junction with the main line at Frieze Hill which the tithe map and John Woods's town map—both drawn in 1840 and showing the cut—also indicate, would have made navigation difficult if not impossible.

Although, in spite of delays, the 'revival of the undertaking' was reported as having 'proceeded satisfactorily, with regularity and without disappointment',[2] there were obstacles to be overcome. One of these arose from persistent opposition by the Trustees of Turnpike Roads at Taunton to the passage of the canal across two roads (on the length of the deviation) which the trustees contended had not been expressly authorised by the Act of Parliament, involving both delay to the work and the danger of expensive litigation. It appeared that whenever the trustees at their monthly meeting met in small numbers they succeeded in recommending and adopting some form of interference which a subsequent special meeting always rescinded and set aside. A solution was reached by the committee, who reported to the General Assembly meeting at the City of London Tavern on 28 June 1832, that they 'have uniformly endeavoured to consult the wishes of the larger body of Trustees, and have at length arranged and executed a special agreement for this purpose, assigning to them sufficient securities for its proper performance, and now flatter themselves that no further difficulty can arise from that quarter'.[3] The securities mentioned, which were deposited with J. Buncum, the trustees' treasurer, appear from accounts to have consisted of a £500 Exchequer Bill valued at £505; and stock of total value £310.

At the same general assembly a report was made on shares with calls in arrears. As had been noted at the committee meeting of the previous day, these, which amounted to 360, had

Grand Western Canal;

360 FORFEITED SHARES,

(By Order of the Committee.)

SHARES IN NAVIGATIONS,

Covent Garden Theatre, Gas Companies, &c.

The Particulars

OF

SHARES

IN THE

Grand Western Canal;

IN THE

Grand TRUNK, LANCASTER, STRATFORD, THAMES, and SEVERN, and other NAVIGATIONS;

IN

COVENT GARDEN THEATRE,

CITY of LONDON and WESTMINSTER GAS COMPANIES,

AND

London University;

WHICH WILL BE SOLD BY AUCTION,

By Mr. SCOTT,

(Nephew of the late Mr. T. SCOTT,)

At the MART, opposite the BANK,

On FRIDAY, the 2d Day of MARCH, 1832,

At Twelve O'Clock, in Lots,

(IF NOT PREVIOUSLY DISPOSED OF.)

Printed Particulars may be had at the *Mart*, and of *Mr.* SCOTT, *Estate and Canal Agent*, 8, CAREY STREET, LINCOLNS INN.

N. B. On the same day will be sold, a *Freehold House*, and *Premises*, producing £40. a Year, well secured, situate at *Clerkenwell*, near the intended *New Street* from *Holborn Bridge*.

FIGURE 2. Notice of sale of shares, 1832

previously been declared forfeited and had been offered for public auction in London on 2 March, the buyer to pay £21 for calls due, but there were no bids. They therefore remained at the disposal of the company, and since it was not convenient to cancel them, as the sum they were expected to produce would be required for the future progress and completion of the canal, and because the Act required that appropriation of such shares should be made with regard to the equal and mutual benefit of all the shareholders—all of whom had a direct interest in rendering the funds effective for the completion of the canal and the consequent acquisition of compensating revenue—it was agreed to divide these shares, as well as some which had lapsed earlier, between all the proprietors in proportion to their existing interests, 1 for every 5 held, on payment of £21 for each.

The committee at this time repeated its conviction that the expenditure was not likely to exceed the calculation of £65,000 and that the revenue to be derived from the 24 miles of canal 'passing through so fine a portion of country' would afford an ample remuneration. In its estimate of possible future revenue the committee noted that it appeared that 40,000–50,000 tons of coal, culm and heavy goods passed westwards from Taunton towards Tiverton and the surrounding country, and that the gradual increase of population combined with the facility and cheapness of carriage offered by the canal communication was likely to augment the quantity very materially. It reckoned:

Assuming therefore at least 40,000 tons, paying average 3s (15p) each, £6,000:

The trade in Lime Westward now amounts to	£1,000
Additional trade, Eastward, Lime, Builders' stone etc	£1,000
Sundries, Grocery, Merchandise etc.	£ 500
Total estimated annual revenue	£8,500

which would, after payment of all expenses, allow a fair expectation of dividends of £2 or more annually on 3,095

FIGURE 3. The canal near Tiverton, about 1833

shares. The total figure quoted was in fact around £1,000 less than those estimated in 1817 and 1819, but the committee pointed out that it was being cautious, saying that although present prospects were still more favourable it was anxious not to overstate the probable amount of revenue.

PROGRESS REPORT

By the middle months of 1832 a considerable portion of the work had been done, as Green reported in a letter to the committee dated 21 June.[4] The line was divided into eight sections or lots, numbered from 1 at the connection with the basin of the Bridgwater & Taunton Canal to 8 where it joined the summit level at Lowdwells. Contracts for the execution of lots 1, 2 and 3 and for 6, 7 and 8 had been offered for public competition and taken at terms considerably less than the estimates, while those for the middle lots, 4 and 5, had been 'delayed from considerations of prudence', but had also recently been contracted for at sums less than those estimated. Lots 1, 2 and 3—those nearest Taunton, totalling 5 miles, 1 furlong, 4 chains in length—were being constructed by Messrs Houghton & Co, the firm which had also contracted for but not yet commenced the 3 miles and 6 chains of lots 4 and 5; lots 6, 7 and 8, measuring 4 miles and 3 chains, were in the hands of H. MacIntosh (probably Hugh McIntosh, who had done much canal work in Yorkshire and elsewhere).

Green noted that in the earth works of lots 1, 2 and 3 little more than half a mile of cutting remained to be executed; boating was required for this further work, however, which would include completion of the embankments, and progress therefore was expected to be slow and to take at least 5 months. Of building work in these lots only three more culverts had to be laid. The viaduct over the Kingston road, near Taunton, was nearly ready to receive the ironwork, which was already prepared. The aqueduct over the Staplegrove stream (and small arches under embankments near this stream) had been completed as also had the aqueduct over the Hillfarrence brook, and a bridge in the Minehead road near Taunton was also finished. The other road bridges in these lots, at Staplegrove, Allerford, Nightingale Green and Trefusis Farm, were ready to receive their ironwork which was prepared and shortly to be fixed. Of the several occupation bridges two swivel bridges were finished and the masonry was ready to take the ironwork and timber of the remainder, all of which was prepared and ready for fixture. Of the three perpendicular lifts to be sited in these lots one, near Allerford, had been commenced and the masonry built to the height of top bank level, while the foundations of the Taunton lift were nearly executed and the masonry expected to be started in about a week.

In lots 6, 7 and 8 the earth work was nearly finished, only 7 chains of cutting remaining to be excavated and the embankments made good—work which it was expected might be done in the course of the ensuing month. Of the building work here all the culverts had been laid and the iron aqueducts at Wellisford and the Ramsey road and the road bridges at Harpford, Cothay Farm and Greenham, as well as the lock at Lowdwells mill were all finished. Of the occupation bridges three were finished and one ready to receive the timber and ironwork, which was prepared for fixing. The masonry of the perpendicular lift near Greenham had reached to about 15ft above top bank level and was expected to be ready for its ironwork in 3 months at the latest, by which time the remaining bridge below the lift might be completed. The inclined plane at Wellisford was sufficiently advanced to ensure its completion at the same time as the Greenham lift, the heading to discharge the water having been driven 160 yards in length and the shafts sunk for the bucket pit.

The ironwork for the various lifts was sufficiently prepared to ensure its erection as soon as the masonry was ready to receive it and Green concluded: 'it may be right for me to state that as far as the several works have proceeded, they have been done in a very efficient manner'.

CANAL LIFTS IN GENERAL, AND JAMES GREEN'S IN PARTICULAR

At this point it may be convenient to consider in some detail what precedents existed for the use of lifts to raise canal boats, and how those which James Green devised for the Grand Western Canal were constructed and worked.

The idea of lifts was not a new one. In a paper which he contributed a few years later to the *Transactions* of the Institution of Civil Engineers (Volume 2, 1838), entitled 'Description of the Perpendicular Lifts for passing boats from one level of canal to another, as erected on the Grand Western Canal' Green himself wrote:

> The merit of the first idea of passing boats from one pond of canal to another on this principle, is justly due to the late Dr James Anderson, of Edinburgh, who published a paper on the subject in his Agricultural Survey of the County of Aberdeen, about the year 1796; but it will be seen on a perusal of that able paper, that the details by which the principle was to be carried out, were left much to the practical man.

There were however other people at this time devising plans for lifts, and even before the date of Anderson's paper. The idea had been considered by the third Earl Stanhope, a man of inventions and a promoter of the Bude Canal, who in the mid-1790s had described plans he had developed for the vertical lifting of boats in correspondence with Robert Fulton.[5] If Green was aware of Lord Stanhope's ideas, as he might have been since he was engineer for the Bude Canal's eventual construction and could have had access to preliminary correspondence, he apparently did not reveal it.

But it is quite probable that Green knew of certain lifts which had been actually constructed and could possibly have even seen one or two of them under trial. The first was one built by Robert Weldon at Combe Hay on the Somersetshire Coal Canal. For this canal, consideration had been given in 1794 to a suggestion from Dr Anderson for the use of his type of lift, comprising two counterbalanced caissons held by chains passing over pulleys between them, with lifting effected by discharging water from the lower container. This however was not really suitable for the particular purpose as it was designed for boats of not more than 20 tons, and preferably of only 10 or 15, and the use of narrow boats was here intended. Alternatives, in the form of inclined planes and railroads, were considered and rejected, and in 1795 the Somersetshire committee decided to experiment with another lift idea, that of Weldon, which had been put forward in 1794. Weldon's lift, described by him as a 'hydrostatick lock' and later called a caisson lock, was designed for carrying a full-length narrow boat and was probably based on an idea for a lift devised by Erasmus Darwin in about 1779. It consisted of a single watertight wooden caisson, capable of taking a narrow boat loaded with up to 30 tons, which could be raised and lowered by rack and pinion gearing within a brick chamber full of water; the caisson was totally immersed in the water, which thus took the weight. Full height of the construction was 88ft and the lift obtained 46ft. Building of the lift started at the end of 1796 and continued through the next year, and early in 1798 the mechanism was given a full test. After an initial mishap caused by the breaking of part of the rack and pinion system during the test, the lift apparently began working satisfactorily and the company was even considering building two more, but then persistent failures developed and the weight of water behind the lift caused the walls of the chamber to bulge. Tenders for rebuilding were invited but none were received and eventually an inclined plane, over which containers were conveyed on rails, was substituted.[6]

Also during the 1790s trials were being carried out with a lift at Ruabon on the Ellesmere Canal in Shropshire, based on an idea patented in early 1794 by Edward Rowland, one of that canal's shareholders, and Exuperius Pickering, a land and colliery owner. By 1796 a successful trial had been made. The construction consisted of a well of water containing a float on which were a number of iron pillars supporting a tank of water which accommodated the boat; addition or removal of water in the chamber caused the necessary raising or lowering of the tank which could be controlled by capstan or rack-and-pinion. It did not survive long, Jessop and Rennie reporting that though it worked well under favourable conditions the lift was not sufficiently robust for daily use.[7]

Back in the westcountry, the use of vertical lifts instead of locks in the construction of the Dorset & Somerset Canal, projected in 1792 to link Poole and Bristol, was receiving the attention of the ironmaster James Fussell. He probably knew of Dr James Anderson's proposals, the Ruabon experiment, and also of an earlier invention of 1790, of a counterbalanced lift, by the Ellesmere Canal's engineer John Duncombe, which was apparently not proceeded with. He was doubtless aware too of the Combe Hay lift under trial not far away on the Somersetshire Coal Canal. Fussell built the Mells lift at the top of Burrow Hill on the general lines of Dr Anderson's ideas, and tests were made in 1800. Though there were certain technical differences, it worked on the same basic principle as those which Green later constructed on the Grand Western; 10-ton boats were raised a height of 20ft by means of two counterbalanced caissons in which the boats were floated. The Mells lift apparently worked satisfactorily during the tests and the company decided to build five more, but money ran out before the canal could be completed, and the works were abandoned.[8]

Another lift in the Midlands was one invented by John Woodhouse for the Worcester & Birmingham Canal. The canal company was anxious to reduce the number of locks needed and the water requirement of the projected canal and agreed to back Woodhouse's experimental construction. This lift, at Tardebigge, was built by 1808 though not tested seriously until early in 1810; it consisted of a wooden tank, large enough to take a narrow boat and weighing 64 tons when filled with water, counterbalanced by a platform weighted with bricks with which it connected by 8 chains, each passing over an individual cast-iron wheel above the tank. Gates at each end could be raised to enable boats to enter and leave, and lifting was accomplished by hand-operated winch. Test results appeared to be reasonably satisfactory, but the company really wanted locks, though a group of proprietors doubted the sufficiency of water and even Jessop advocated a lift in preference. After further trials, however, the company consulted Rennie, who in his report expressed the view that the lift was too complex and delicate to withstand general use on a canal. Subsequently it was estimated that water would after all be adequate for locks, one of which was built to replace the lift in 1815.[9]

A yet further experimental lift was constructed between 1812 and 1815 on the Regent's Canal at Camden Town. This, designed by Colonel, later Sir William, Congreve, was to operate on a hydro-pneumatic principle, with two caissons linked by chains passing over overhead wheels. The two chambers connected at the bottom, and compressed air trapped beneath the water, passing from one chamber to the other, assisted the hand operation of the lift. But the system could not be made to work efficiently and after spending much money on the scheme the company dispensed with it and reverted to locks.[10]

No subsequent canal lifts were constructed until James Green's use of them on the Grand Western. Fifteen years therefore had elapsed since the abandonment of the last of the series of prototypes—none of which had survived into established use—when Green suggested them in his report of 1830. He must, it appears, have had faith in his own ability to succeed where others had failed and to have been convinced of their suitability for negotiating the changes of level which the canal he was engineering demanded.

The seven lifts of the Grand Western Canal, when they were eventually completed after much difficulty, each consisted of a pair of caissons, suspended from carrying wheels and containing water, into which the boats were floated, power being provided by a preponderating quantity of water added to the descending caisson. In his paper,[11] which, it will be noted, was not published until 1838, by which time the theories he expounded in it had been found not wholly workable without modification, Green explained that his lifts had been designed for a particular situation, where a considerable ascent had to be overcome within a short distance and where both water and funds were insufficient for locks. The trade expected was chiefly the carriage of coal, culm for burning lime, and limestone, conveyed in trains of 4, 6 or 8 small boats linked together, each train being drawn by one horse. With economy in cost the main object, trade would move slowly and convey a large quantity. To prevent hindrance by too frequent detaching of boats for their single passage over the lifts, it was desirable therefore that the pounds of canal between the lifts should be as long and the lifts as few as possible. The boats, he noted, were built to carry 8 tons each, 26ft long and 6½ft wide, and drawing when laden 2ft 3in of water so that a canal 3ft deep was sufficient for their use.

Green described the lift of his invention as consisting of two chambers, similar to those of a lock, with a pier of masonry between them. Each chamber accommodated a wooden cradle or cistern, fitted with water-tight lifting doors or gates at each end, in which a boat could be floated. The side walls of the chamber and of the pier were carried up from the foundation below the bottom level of the lower canal to the top bank level of the higher pound, the perpendicular height of the actual lift equalling the difference in the levels of the two pounds of canal. Longitudinal and transverse arches in the centre pier, and platforms at the higher ends of the chamber, gave access for adjustments to the cradle, while the transverse arches and others in the retaining walls of the structure lessened the mass of masonry and admitted daylight. Water from both the lower and upper pounds of canal was prevented from flowing into the lift chambers by lift-up gates or doors.

Drawing of perpendicular lift, reproduced from James Green's 'Description of the Perpendicular Lifts for passing boats from one level of canal to another, as erected on the Grand Western Canal' in the Transactions of the Institution of Civil Engineers, Volume 2, 1838.

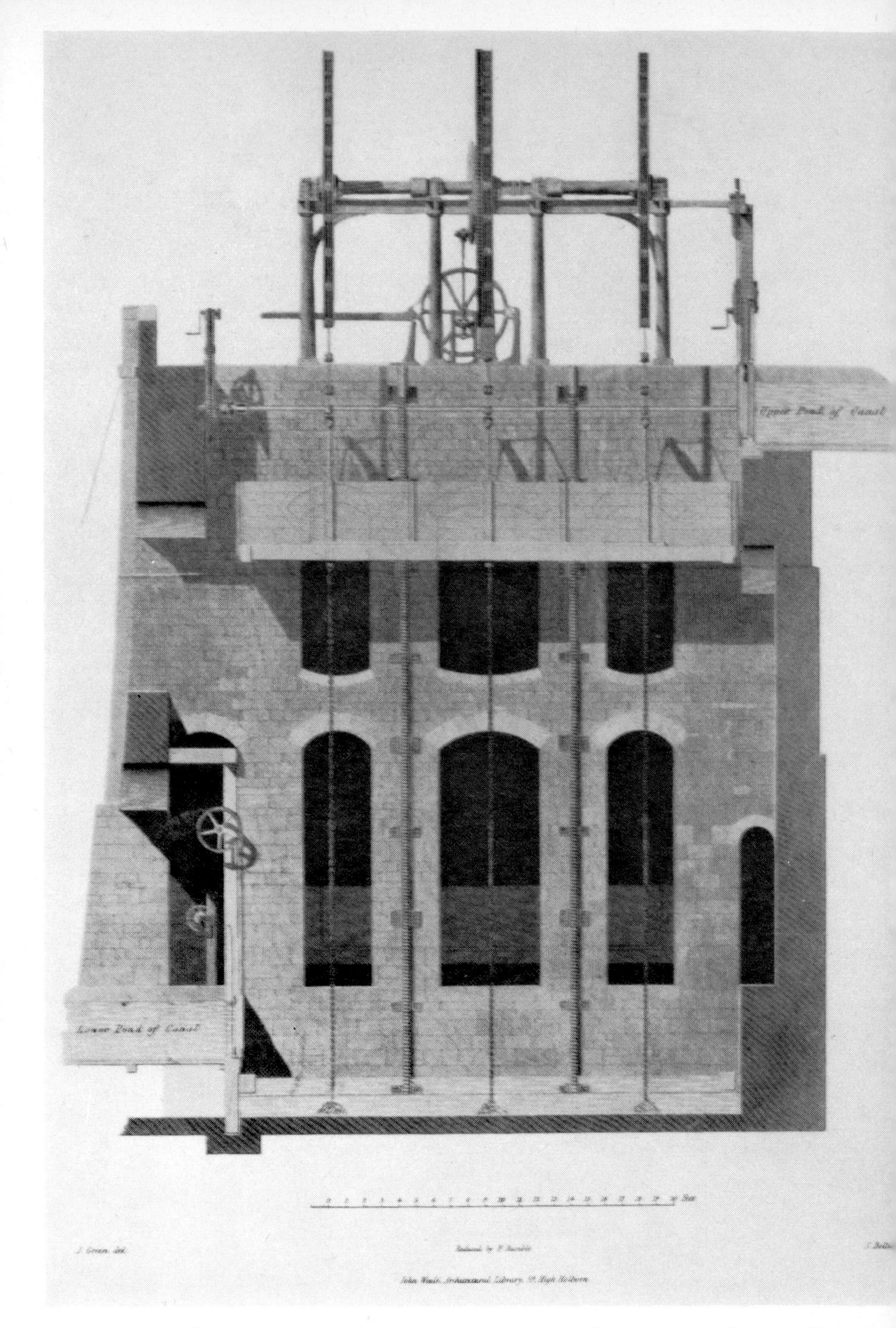

Page 90 Drawing of perpendicular lift, reproduced from James Green's 'Description of the Perpendicular Lifts for passing boats from one level of canal to another, as erected on the Grand Western Canal' in the Transactions of the Institution of Civil Engineers, Volume 2, 1838.

The floors of the chambers were sufficiently below the bottom of the lower canal to accommodate the coil of the balancing chain underneath the cradles and to leave the cross beams of timber—which supported the cradles when at the lower level—clear of water, a drain being laid from each chamber to prevent water accumulating above the height of the bottom of the canal. The sides of the cradles were well secured by wrought iron knees on the inside, riveted to wrought iron straps on the outside, and the ends of the cradles had cast iron frames bolted to the timber, all joints being caulked and made watertight.

On the top of the walls of the lift a framing of cast iron was erected, consisting of 10 upright hollow columns, 9ft high and of 12in diameter, secured to the masonry by strong wrought iron holding down bolts. The columns were braced together by lateral and transverse beams, the framing supporting a longitudinal cast iron shaft, 22ft in length and of 10in diameter, with the bearings seated on brasses. On this shaft were fixed three cast iron wheels or pulleys, 16ft in diameter, for carrying the wrought iron chains which supported the cradles, the opposite points in the wheels' diameters being directly over the centre of each chamber of the lift. The two outer pulleys simply carried the chains, but the centre one was connected through gearing to a handle fixed on the side of each chamber wall. This could be used to work the lift when necessary—probably to make small adjustments to bring the caisson exactly opposite the canal entrance. To this hand gear was attached a cast iron brake wheel and brake lever for regulating the speed of ascent and descent of the cradles when, as was generally the case, a preponderating weight of water in the cradles provided the motive power.

The chains used to support the cradles were of the best wrought iron and bar chain, with coupling or connecting joints and steel pins, and the wheels and pulleys being cast to allow proper seating of the joints to prevent slipping of the chains. The cradles were attached to the chains by means of strong wrought iron suspension bars fixed to each side in the direction of the three pulleys. These suspension bars were connected in pairs by a cast iron beam across the cradles, at sufficient height above the sides to allow the boat to pass under them. A round wrought iron bolt, 3in in diameter, was attached to each chain, and passed through a fitted aperture in the centre of the suspending cross beams. On these bolts was a strong square threaded worm screw, and underneath the beams a strong brass nut, by means of which the cradles could be adjusted to their proper horizontal position. There were also wrought iron bridle bars with nuts and screws placed diagonally from the ends of the cross beams to a link in the suspending chains, by which the transverse level or position of the cradle was adjusted.

The length of the suspending chains was so arranged that when one cradle was at its proper level at the bottom of the lift the other was at its proper level at the top. With each cradle containing the same quantity of water, weights would be equal but for the difference in the length of chain between the cradle at the bottom and the one at the top; to remedy this and to preserve equilibrium, chains of an equal weight per foot to the suspending chains were attached to the bottom of each cradle, with one end resting on the floor of the chamber, gathering there or elongating as the cradle above it was lowered or raised. No more was needed to put the machinery in motion than a power equal to its *vis* inertia and friction, together with the required velocity. To achieve this the chains were so adjusted that when one cradle was at the bottom of the lift on the proper level to receive a boat, the cradle in the opposite chamber was not quite up to the level necessary to receive a boat from the upper pound of canal; the difference found necessary in operation was not quite 2 inches, producing in the cradle a preponderating weight of only 1 ton, which could be regulated as required.

The cradles were so suspended that when raised to their proper height they came to within half an inch of the higher stop gate to the canal. The cradle at this point was then forced forwards by means of a forcing bar of cast iron at the rear, which was moved by a hand multiplied gear so that a turn or two of the winch would force the cradle against the framing of the stop gate and no water could escape between them. The stop gate of the higher canal level was lifted by a winch gear on the sides of the chamber which put in motion chains passing over pulleys fixed in a transverse timber framework. The operation was said to be easily and speedily performed, and, by a strong cast iron bolt fixed on the higher side of the stop gate being moved (by a man's foot) into a corresponding square staple in the door of the cradle, both doors could be raised at the same time. Lifting of the doors caused an immediate flow of water from the canal into the cradle, which was sufficient to allow the boat to pass into the canal at its proper level, and, when the doors were let down, provided the necessary preponderance of water in the cradle to produce its descent, which commenced on the winding back of the forcing bar.

The descending cradle, on arrival at the surface level of the water in the lower pound, came in contact with two inverted wedges of wrought iron fixed at the back or higher end of the lower chamber, and by sliding against them in its further descent the cradle was forced tightly against the inner side of the framing to the lower stop gate, the meeting becoming watertight. The door at the lower end of the cradle had two inverted half staples of iron fixed at the top which fitted into iron mortars in the stop gate when the cradle had arrived at its proper level, and thus by raising the stop gate in a similar way to that on the higher level, the door of the cradle was also raised and the boats could pass freely from and into the cradle. Lowering of the gates was of course done in the same way.

Green emphasised that the essential requirement for this particular construction was strength of material and proper attention to the arrangement of the several parts. The principle was simply that of equally poised suspended weights, and it mattered little to what height (within reasonable limits) the lift extended, provided the parts were proportionately strong and length and weight of chains, power and braking were modified accordingly. The advantages he enumerated as: firstly, economy in the expense of construction as compared with common locks; secondly, the saving of time in passing boats from one level to another; and thirdly, the small consumption of water compared with common locks.

In the lifts described, Green noted that consumption of water was about 1 ton to 8 tons of cargo raised from one level to the other, whilst the same weight was at the same time correspondingly lowered. Adding to this a similar quantity lost by leakage of the gates (which in practice was very small) the whole consumption was only 2 tons of water to 8 tons of cargo, whilst locks generally required 3 tons to 1 ton of cargo. Thus the saving of water in one system over the other was nearly 92 per cent. Moreover it had been observed that a quantity of water exactly equal in weight to the amount of gross tonnage of boats and cargo would pass either up or down the canal in a direction contrary to that of the load, so that if the trade was all downwards and boats returned empty, water equal in weight to the whole of the loading being passed down would pass up from the lower level of canal to the highest, and vice versa, independently of the use or waste in working the lifts. The observation applied equally to the difference of balance of tonnage between the up and down trade.

The time taken in passing one boat up and another down a lift 46ft high was found to be 3 minutes, constituting a notable advantage over locks, which required an average of 5 minutes for boats to pass up an 8ft rise, even taking into account that boats using locks usually conveyed 25 tons of cargo at a time.

Following his reference to Dr Anderson already quoted (p 84), Green remarked that the way he had put the principle into practice differed in one essential point very materially from that suggested by Anderson who had proposed: 'that in order to give the upper cradle power to descend, a sufficient quantity of water should be drawn out of the lower cradle to give the required preponderance in the descending one, which quantity of water would pass by the drain from, and be entirely lost to the canal, in addition to the leakage'. As Green pointed out, by his method the cradle having to descend was charged in the first instance with so much more water than was contained in the ascending cradle as would give the required preponderance and that this water, on the passage of the boat from the cradle into the lower pound of canal, was delivered into the lower pound and was therefore available in the use of the several lifts below. However, as we shall see, the matter was more complicated than Green foresaw.

FRUSTRATED PROGRESS, AND THE OUTCOME FOR GREEN

The building of the canal did not proceed as quickly or as smoothly as Green doubtless anticipated when he had presented his 1832 report. In the committee's report to the annual General Assembly in 1833[12] the 'regular and progressive execution of the works' was noted, as also was the receiving of enquiries regarding wharves and boats which augured well for future trade, but the committee was unable to report the degree of completion it had hoped, particularly with regard to the three miles from the town of Taunton which it had been confidently expected would by this time have been navigable. Nevertheless, it was emphasised that, though vexatious, the delays involved 'no imputation of censure upon the zeal, diligence or exertions of either the engineer or the contractors'.

One factor in the delay arose from the decision, in order to secure a uniform level of water, to construct a lock at the foot of the first lift, near the junction with the basin of the Bridgwater & Taunton Canal; this was being built, the masonry was nearly complete and the machinery prepared. From the top of this lift the canal was navigable for nearly a mile, as far as an embankment where progress was slow, due to the necessary earth having to be conveyed a considerable distance on each side by boats and waggons and requiring time to settle securely. The distance remaining to be completed here was about 300yds, and progress was at the rate of only 30yds a week. Another embankment of the same type, farther along in Lot 3, also remained to be completed and was expected to take the same length of time, but, with bridges and culverts being already generally complete, it was anticipated that by then the whole extent of Lots 1, 2 and 3 would be navigable, a distance of over 5 miles which included three of the lifts. Of these, work on both the one near Norton Brook and that at Allerford, as on the Taunton lift, was in progress. In Lots 4 and 5, comprising the next three miles, satisfactory progress had been made despite difficulties in arranging the line, involving consultations with landowners which had caused delay over possession of a considerable portion; the large embankment in Nynehead Park and a proportionate quantity of deep cutting were the principal obstacles to be overcome, together with completion of the three lifts in the section—those at Trefusis, Nynehead and Winsbeer—on which progress was said to be in a forward state. In Lots 6, 7 and 8, constituting the remaining 4 miles, the whole of the canal itself was navigable and the bridges and culverts complete, though continuous navigation was obstructed by the inclined plane at Wellisford, said to be 'proceeding with rapidity' and the lift at Greenham, the masonry of which was nearly finished and the machinery ready.

The committee's report of the following year, 1834,[13] was also disappointing. Apologies were expressed for the delay which, it was stated, was not due to any one cause but rather to an accumulation of petty difficulties which, through inactivity and lack of foresight, the contractors had failed to overcome, despite repeated remonstrances by the committee and engineer. With some difficulty the contractors had, however, been at length roused to exertion and had signed an agreement under heavy penalties to complete the first three lots as far as Bradford by 1 August and the whole line by the following

1 October. It was also reported that Green had found it necessary to overspend on contracts; the estimates for these had been £44,555, but the amount actually paid and due had reached £49,544, the excess being accounted for by:

	£
2 locks, one at each end	1,000
Several additional bridges and buildings necessary by agreement with landowners	1,500
Expense of raising banks of canal, widening bridges, altering aqueduct, improvements of the road and other works consequent upon opposition of Turnpike Trustees	1,500
Implements and establishment at Yard at Taunton, tramplates and wharfs	1,000
	£5,000

The report continued:

> The actual increase of expenditure is to be attributed first to the change of line from the Parliamentary to the Northern Substitute which was essential to our prosperity; and consequent necessity of purchasing more land at a high price, and second to being compelled by various untoward circumstances to pay throughout the line much more than the real value of the land, exceeding altogether the original calculation more than £9,000, but it was hoped and even anticipated that a sum nearly to that amount might have been saved on the contracts, the fact however has proved otherwise.

It was necessary to borrow £12,000 and to pay interest out of the revenue from the canal's summit level. Consequently, £9,000 was shortly secured in £1,000 mortgages subscribed by various individuals, and in 1836 another £3,000 was raised by the same means. Subsequent loans, including mortgages by committee members, raised the total sum borrowed at this stage to £14,750.

There was some response by the contractors, for on 5 February the *Taunton Courier* reported:

> Seven barges laden with coal attended by another barge in which were a band of musicians and several spectators, yesterday passed through the aqueduct which crosses the Kingston road, having been brought to that level by the novel and efficacious process of lifting barges, and proceeded to Bradford where, a wharf having been opened, the nuisance of the coal carts along the Wellington Road may be expected will in a great measure now cease, and entirely go when navigation is further completed Westward which it will be in a few months.

But the 'few months' elapsed and still the whole length was not opened, and not, it appeared, entirely due to the procrastinations of the contractors. When the committee reported on 25 June 1835[14] it was again with regrets at the delay and the explanation:

> the true cause appears to rest in the novelty of the plan of the lifts, and the want of foresight as to actual inconvenience which might arise in using them. Although in January 1834 the lift near Taunton was reported to be ready and to answer every purpose, Mr Green relying with too much confidence on theoretical principles never subjected it to a full and fair trial, so that many practical difficulties were only gradually developed and detected, which had not in the first instance been either observed or anticipated: as other lifts came into operation, several serious obstacles to the free and perfect action were discovered and various expedients applied to remedy them without effect, till at length a secondary lock or chamber was continued or applied, besides improved arrangements of the machinery, all of which required time for their execution. The Engineer having engaged to complete these lifts for a certain sum, all these experiments alterations and additions, have been made at his expense, but at a serious injury to the Company, for the delay and for the continued cost of superintendence without additional revenue and to Mr Green at a heavy pecuniary sacrifice. That all these impediments are surmounted in a satisfactory manner the Committee are willing to believe, an opinion confirmed to a certain point by the opening of the Trade, including five lifts, as far as Wellington.

Regarding rates, the report pointed out that the cost of land carriage of coal from Taunton was incredibly low and that there was no hope of creating a trade on the canal unless rates could be fixed which would preclude all competition. It was suggested that coal and merchandise should be carried for 2d per ton per mile, and stone for 1d, with for the first year a discount of 25 per cent on all coal not carried beyond Wellington.

The report continued:

> there is no reason to anticipate less amount of revenue arising from this source than had been formerly calculated upon, while there is good ground to expect a much larger return Trade downwards in lime and agricultural produce than had ever been anticipated and there exists also great probability that a considerable revenue may be derived from conveyance of building materials. For the accommodation of the expected trade some expense must be incurred at Tiverton, for the Wharf, wall etc., to the extent probably of £1,000, for which there will be ample remuneration in rent, and eventually a house must be built at each Lift to ensure the protection and proper use of machinery, at an expense probably to the same amount.

As one year proceeded to the next and through navigation was still not possible due to faults in the lifts and the non-functioning inclined plane, and with its financial resources diminishing, the Grand Western Canal Company became increasingly dissatisfied with its engineer, whom it dismissed early in 1836. How the break came, and how bitter whatever quarrel there was is not known, but a brief notice in the *Taunton Courier* of 3 February which read 'Notice is hereby given that Mr James Green has ceased to be Engineer to the Company' signed by Fred Leigh, principal clerk, and dated 27 January 1836, communicated the information plainly and simply. John Twisden, a retired Royal Navy captain and a man it would appear of some financial means, who had been associated with the canal construction at least since 1829, drawing a salary of about £140 per annum, was retained to superintend the work, and James Easton of Taunton was asked to survey the machinery.

PROVIS'S REPORT

In May 1836 the committee requested the engineer W. A. Provis to visit the canal works from Taunton to the summit level and to investigate the lifts and inclined plane and the causes of the plane's failure. This he shortly did, travelling from London to Somerset for the purpose within the month, and reporting to the committee on 28 June.

Provis's report, clear and concise, provides the best description that is available of the Taunton–Lowdwells length of the canal as it was constructed. It reveals the modifications which Green had had to resort to in order to make his lifts work, describes the intended method of operation of the inclined plane and gives details of water usage.

Provis commenced his investigations and observations at the canal's Taunton end, at the regulating lock near the junction with the Bridgwater & Taunton. The gates of the lock, which also served as stop gates, were kept open or closed depending on the relative levels of the water in the two canals, and construction of the lock, over which there was a bridge, he found generally good. Regarding bridges on the canal a few general observations were given. The fixed road bridges Provis found were composed principally of masonry with, in most cases, cast iron ribs to carry the roadways across the canal; where there was enough height to allow arches over the water these were sometimes made entirely of stone and sometimes mixed with bricks, and where it had been necessary to keep the road over the canal very low the towpath was sunk under the bridge to a lower level than the canal and protected from the encroachment of the water by cast iron plates. The fixed foot bridges had stone abutments with cast iron plates screwed together to form the pathway over the canal, and the swivel bridges were of timber with platforms and wings of masonry. Nearly all the stone copings of the towpath and water wing walls were found of too small dimensions and often of bad quality with inadequate bonding, but in other respects the bridges appeared well designed and substantially constructed.

The aqueduct which conveyed the canal over the road from Taunton to Kingston had a cast iron trunk for the waterway, and ribs of similar metal on each side to carry the towpaths, supported by abutments and wing walls of masonry. Apart from some bad bonding and inadequate coping the essential work was said to be excellent. Other aqueducts which carried the canal across roads, public and private, were either wholly of masonry or of masonry and brick, in most cases with a cast iron trough for the waterway; generally the masonry was of coarse rubble work, plain and substantial, though the aqueduct over the drive to Nynehead Court was of neat and ornamental ashlar work. Four aqueducts carried the canal across water: over the Norton Brook (cast iron trough and ribs, spanning 30ft and resting on ashlar masonry), the Hillfarrence Brook (skewed, flat and close to the water), the River Tone (iron troughs with stone arches carrying the paths and abutments of masonry, of 30ft span) and the stream leading to Fox's factory near Wellington (an iron trough with brick arch and masonry). Culverts passing under the canal for drainage or irrigation were found generally rough but sufficient for the purpose; at Greenham Barton road bridge and at Greenham bridge water for irrigation was taken over the canal in cast iron trunks.

The bulging of retaining walls in certain places was noted and in several instances the towpath required drains to carry off land springs. Provis also reported that he was informed certain pounds of the canal lost water to a considerable extent, particularly below Allerford lift where, when the river was very low, the water was said to sink 2–3in in 12 hours; Edward Holt, who had helped to make this part of the canal, told him that there was no lining through the gravelly deep cutting below the lift. There were public wharves at Taunton, Bradford and Payton, belonging to the company, and others which were private property at Bradford, Tonedale and Wellisford, which Provis considered were sufficient for current requirements.

Discussing the lifts, Provis explained that the first three, Taunton, Norton and Allerford, were of the same general form and character and might be classed together. They were originally constructed on Dr Anderson's principle except that no means were provided for keeping the water in the caisson chambers at a lower level than the water in the canal. The descending caisson could not therefore by itself sink deep enough to allow the boat it contained to float into the lower pound and the ascending caisson to rise high enough for its boat to pass into the upper one. Attempts to remedy this had been made by erecting gates to cut off communication between the lower pound of the canal and the space under the caisson, and pipes laid to drain off water from the thus enclosed chambers, but the pipes had proved inadequate and instead locks were constructed, a pair at each lift, at the lower end of the caisson chambers. The water level in the lock was raised by a supply from above until high enough for the boat to be floated in, and then the boat was let down the 3ft to the lower pound in the same way as with a common lock. Both the doors of the caissons and the gates of the chambers were of wood and iron, the former falling flat on the bottom of the caissons, and the latter moving horizontally on pivots. The gates at the lower ends of the locks were of timber, and moved horizontally.

Apart from the functional problem and the necessity for the locks to correct it, Provis was informed by people employed on the canal that these three lifts, which had been in use for two years—since 1834—had worked reasonably well and that accidents had been few and trifling. At the Taunton lift there had been no accident until the previous April, when, due to the carelessness of a boatman who was assisting, the water was drawn from the lower caisson while the upper one remained loaded, so that the loaded caisson descended too rapidly and was broken. At the Norton lift early in the year the main shaft had broken and one of the caissons was damaged, while at

Allerford apparently the only incident had been the breaking of the pinion working in the large wheel on the centre of the main shaft. Provis noted the single brakes on the lifts for checking and regulating the descent of the caissons and stated that there should be two. He saw a loaded boat passed up and another down the 12½ft Norton lift while he was there, and observed that the time occupied from entering the lock at the lower level to floating into the upper pound was 8½ minutes.

The Trefusis lift differed from those already described in having the gates of the caisson chambers and of the caissons lifting vertically, and balanced and ingeniously contrived so that both were lifted by the same machinery at the same time. Consequently the locks could be shorter and the water consumption therefore diminished. As the lift was not finished quite as soon as the previous three the locks were added without the intermediate expedient of the drainage pipes. Provis was told that an accident had happened here early in 1836 involving the breaking of the main shaft; he noted that there was now a double brake at this lift.

The Nynehead lift, except for having only a single brake, was similar in construction to Trefusis. The masonry at the upper end, having been built on an artificial embankment, had sunk, and raising was required when the earth below had become consolidated, but nothing amiss had occurred except for a few cogs of one of the pinions breaking when the lift was first set to work in June 1835. Very similar was the Winsbeer lift which had also been worked since the summer of 1835; the only accident having been the breaking of a bevel wheel which worked in connection with the lift's single brake.

The last lift, that at Greenham, was the only one in which Dr Anderson's principle had been fully carried into effect, a drain having been brought into the bottom of the caisson chamber to draw off its water to a sufficient depth and the necessity for locks at the lower end being consequently eliminated. Pressure of the high earth embankment had forced the side walls of the caisson chambers several inches inwards, and the front wall out, but there had been rebuilding and reinforcing, and no sign of recent movement was apparent.

At Lowdwells, where the new work joined the earlier-built summit level, there was a lock with a rise of about 3½ft, capable of passing four of the small canal boats at once. It was considered by Provis to be an 'excellent piece of workmanship' but the possibility of raising the Greenham lift and the canal banks above it, in order to obviate the Lowdwells lock and to use it solely as a stop gate, was being considered.

The Wellisford inclined plane occupied a considerable part of Provis's attention. By the plane an ascent of approximately 80ft in a distance of 440ft was achieved, over a gradient of about 1 in 5½. Two parallel lines of railway were laid along the slope and on these the canal boats were carried, travelling on special cradles fitted with wheels. At each extremity of the plane the railways passed into and along the bottom of small docks which were deeper than the canal, so that a cradle with its wheels on the railway was not higher than the canal bottom, in order that a boat could be floated on to it. An endless chain passed around a large wheel moving horizontally on a pivot between the two small docks at the bottom and also around another, similarly sited, at the top, the wheels' diameter being such that the lengths of chain were kept vertically over the centres of the two lines of railway. Each cradle had an upright frame of iron on each side and a connecting cross bar at the top, high enough above the platform to enable boats to float underneath. A short buckling chain from the cross bar attached the cradle to the endless chain. With one cradle in the dock at the bottom of one line, the two boats could be floated to their positions over the two cradles, the chain put in motion and the cradles dragged forward, to rise out of the docks and lift and carry the boats forward on their respective inclines.

In order to confine the water of the upper canal its end embankment was raised above the water's surface. Ascent of this was managed by a short inclined plane rising a little above the water's surface up which the boat had to be pulled before descending the principal plane. The machinery was so arranged that the boat about to descend was pulled up the short ascent and over the apex of the two inclines by the time the ascending boat had arrived at the foot of the main plane. Immediately after the descending boat had passed the apex its gravity aided the machinery in pulling up the ascending boat, and by the time it reached the bottom and was afloat again the ascending boat had reached the apex, from which its own gravity carried it down into the upper canal.

Clearly, if the ascending and descending weights were always equal, no greater power would have been required than that sufficient to overcome friction. But in the canal's variable trade there might be considerable weight excesses one way or the other and suitable power was essential. This was supplied by means of two iron buckets suspended in an 80ft-deep well sited at the top of the plane, which were made to rise and fall alternately by the weight of water added to the descending bucket. On reaching the bottom of the well the bucket's gravity forced up a valve which allowed the water to escape by a drain to the canal's lower pound. Then, empty, it would be ready to rise again, as the other bucket, filled from the canal's upper pound, descended and drew up another cradle and boat.

When Provis visited the site a breach had occurred in the bank at the head of the inclined plane and the water was out of that part of the canal, so that he was unable to see the system at work, but he assembled some details from James Easton who had carried out trials a few weeks before.

The buckets used in the well were of a size capable of holding rather more than 10 tons of water. When the endless chain was attached to the machinery without the cradles it had been found that about 4 tons of water were needed in the bucket to put the system in motion, and that when the empty cradles were added nearly a further 4 tons, or a total of practically 8 tons of water were required for motivation. When an empty boat was put on each cradle and the bucket filled to its 10-ton capacity the power was however found insufficient to get the upper cradle to the apex of the plane. Taking into account experiments which had been done by temporarily raising the buckets higher to evaluate the lifting requirements per ton of cargo, and calculating from the figures already arrived at, Provis came to the conclusion that in order to overcome friction of the machinery and the resistance of the cradle, and to lift a boat fully loaded with 8 tons, 25 tons of water would be required—two-and-a-half times the quantity which the existing buckets could contain. This was of course considering the extreme case of a loaded boat being brought up and none sent down, but it was a situation which was often likely to occur, besides the fact that every descending loaded boat had first to be raised up the ascent from the upper canal solely by the power of the loaded bucket. In spite of a suggested minor adjustment in the attachments of the endless chain to the cradle to reduce friction, the power supplied by the system which was in existence was clearly inadequate.

As he had been unable to see the Wellisford machinery in operation Provis travelled farther west to examine the Hobbacott inclined plane of the Bude Canal, which worked on a similar principle. Unfortunately he was again disappointed as he found the great plane being worked by its stand-by steam engine, because, as so often happened, the bucket-in-the-well system had suffered a breakage and had been out of action for some weeks. He was informed, however, that when in good working condition a loaded boat weighing 5½–6 tons was taken up and an empty boat let down in about 3 minutes, at the expense of 15 tons of water in the bucket. No cradles were used at Hobbacott, the Bude Canal boats being fitted with wheels and travelling up the 935ft length of that plane, with its 225ft rise—and the five others—directly on the iron railways.

Considering that 15 tons of water were needed at Hobbacott to raise boats weighing 6 tons, it was quite apparent that the 10 tons of water provided for at Wellisford was far from enough to raise cradles and boats loaded with 8 tons of cargo. And Provis was of the opinion that the existing machinery at

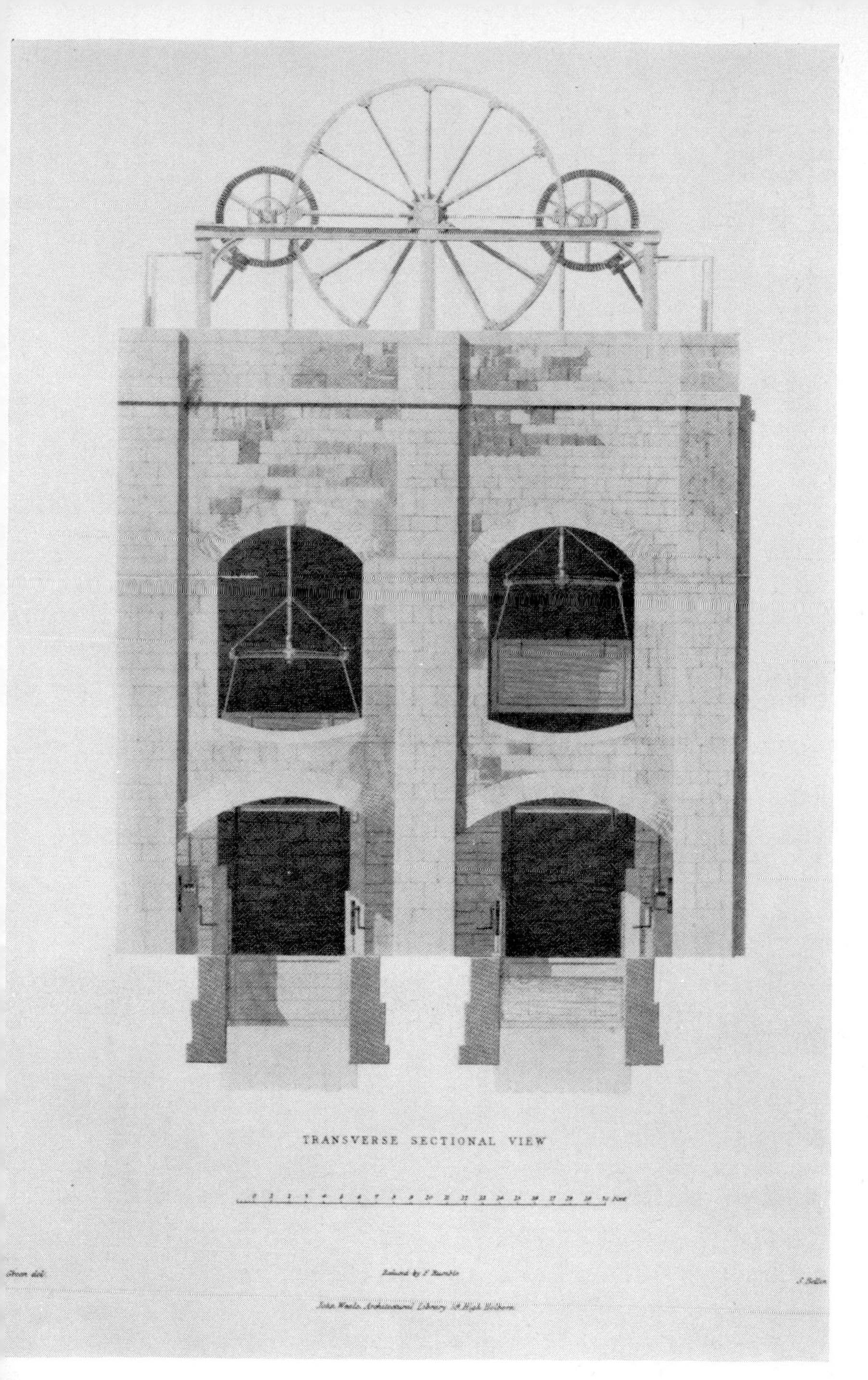

Page 107 Drawing of perpendicular lift, reproduced from James Green's 'Description of the Perpendicular Lifts for passing boats from one level of canal to another, as erected on the Grand Western Canal' in the Transactions of the Institution of Civil Engineers, Volume 2, 1838.

Page 108 (*above*) The fine ashlar arch of the aqueduct by which the canal, crossed the former drive to Nynehead Court; the railway arch is seen in the distance; (*below*) view down the slope of the former Wellisford inclined plane. The boats travelled on wheeled cradles which ran on rails and were raised 81ft in a distance of 440ft.

Wellisford was not sufficiently strong to bear the increased strain of the greatly augmented power required.

The latter part of Provis's report considered the canal's use of water. The water for the length from Lowdwells to Taunton did not come, as had earlier been intended, from reservoirs, but was provided by the same springs near Lowdwells that supplied the summit level. Lowdwells lock received its requirements direct and then the water flowed on, feeding the various lifts in succession and eventually entering the Bridgwater & Taunton Canal. One horse could draw eight boats and this was a convenient number from the point of view of water usage and the operation of Lowdwells lock, which took four boats at a time and consumed 53 cubic feet of water for each ton of cargo when accommodating its full complement on alternate ascent and descent. At Greenham lift the water consumed in operation was estimated at 8⅛ cu ft per ton of cargo and in this case, being discharged by its drain into the river, was totally lost to the canal. Average water requirements per ton of cargo at the Wellisford inclined plane Provis estimated at 56 cu ft, at each of the Winsbeer, Nynehead and Trefusis lifts 53 cu ft, and at each of those at Allerford, Norton and Taunton, which had slightly larger locks due to the caisson chamber gates opening horizontally, 59 cu ft. The six lower lifts were thus seen to require considerably more water for their operation than the lift at Greenham, but with the important difference that the water was not lost to the canal but passed along it. In fact, as the figures show, the operation of the lifts required more water than was sent down by the working of the lock, the sum of the maximum lift requirements and the wastage of Greenham making a total abstraction from the summit level of 67⅛ cu ft of water for every ton of cargo passing either way. Had the lifts been properly constructed, not needing locks, and the water that was used returned to the canal and not wasted as it was at Greenham, the water consumption would have been minimal. In order to obviate waste from leakage Provis recommended lining the canal wherever necessary, and if this were done, and the system of lifts made perfect, he considered it would be desirable to continue the summit level to Greenham lift and use the lock at Lowdwells solely as a protecting stop gate.

COMPLETION

Months of quandary followed the receiving of the Provis report. The Grand Western Company had on its hands a canal which was almost completed and on which a great deal of money had been spent (and borrowed) but which was not yet usable, due mainly to the great obstacle of the inclined plane which would not work. Available funds were used up and receipts for tolls and rents were exceeded by current expenses. Captain Twisden, the superintendent, put his hand into his own pocket and advanced over £1,000, but negotiations in 1837 for a loan from the Exchequer Bill Loan Commissioners were unsuccessful.

An incentive to completion was the obtaining of an Act in 1837 by the Bridgwater & Taunton Canal Company for continuation of its canal from Huntworth to below Bridgwater where there was to be a lock; this would obviate the bore on the River Parrett and enable goods to be transhipped in still water and was expected to help trade on the Grand Western. A party of Grand Western Canal subscribers which included Isaac Cooke, a leading light also in the Bridgwater & Taunton, engaged two engineers, Thomas Maddicks, who had visited canals in Shropshire, and Isaac Whitewood, who had earlier been involved with the Bude Canal and was currently assistant engineer on the Chard, to advise on the canal's completion, and they had considered that the expense necessary at Wellisford was £1,000 and at the Greenham lift £250.

A steam engine was under consideration for working the inclined plane, though Twisden (at this time eighty years of age) thought that it would not answer and submitted a plan to the committee in London for 30 locks at Wellisford 'on a novel principle', to cost £4,000 and to take 18 months to build. The engineers however, felt that £10,000 and 2 years were the more likely figures for such a scheme.[15]

In the event a steam engine was decided on, and, with further loans from committee members and from Twisden, a 12hp engine costing £800 was obtained. The work at Wellisford and at Greenham was completed, and on 28 June 1838 the canal was at last fully opened, the cost of the extension to Taunton having amounted to an approximate £80,000.*

News of the canal's completion must have been received by James Green with some poignancy. How he came to fail in his calculations for the Grand Western Canal's inclined plane, after all his experience of canal construction, is a mystery which will never be solved. Though he remained Devon's county surveyor until 1841 and was consulted on engineering projects elsewhere—on Newport docks, in Bristol and in the building of the South Devon Railway—he did no further work on canals. Green died in 1849.

The only other successful canal lift built in Britain has been the Anderton lift between the River Weaver and the Trent & Mersey Canal, opened in 1875 and still in use, powered now by electricity. A model of it can be seen at the Waterways Museum at Stoke Bruerne. Other lifts are in use in countries abroad.

There is a model of the Wellisford inclined plane at the Somerset County Museum, Taunton.

* Mr Charles Hadfield's figure of £105,000 seems too high.

CHAPTER 5

The Coming of the Railway

WHEN, in 1838, construction of the Grand Western Canal was complete—or as complete as it was ever to be, since fulfilment of the original plan for a channel-to-channel waterway had long been abandoned—it might have been expected that the company's worries were at an end and that a time of financial recovery and prosperity lay ahead. Constructed at a total cost of at least £320,000–£330,000, the 24½-mile Grand Western Canal had made possible, by its junction at Taunton with the Bridgwater & Taunton Canal (from which the southward construction of the Chard Canal was in progress), a through route for water transport from the Bristol Channel to Tiverton. Tub boats carrying 8 tons could now navigate the 13½-mile length from Taunton to Lowdwells, on which the 262ft rise was accomplished by seven lifts and an inclined plane, while barges loaded with up to 40 tons could travel on the broader summit level between Lowdwells and Tiverton. Surely, after all the effort, the company was due for the fulfilment of its trading hopes.

TRADING TRENDS

A marked rising trend in income from tolls indeed began in the late 1830s. In 1836, the year following the partial opening of the Taunton section to Wellington, £1,164 was produced, compared with £971 in 1835. In 1838, when the canal was fully navigable, there was a jump to £2,754, followed by a

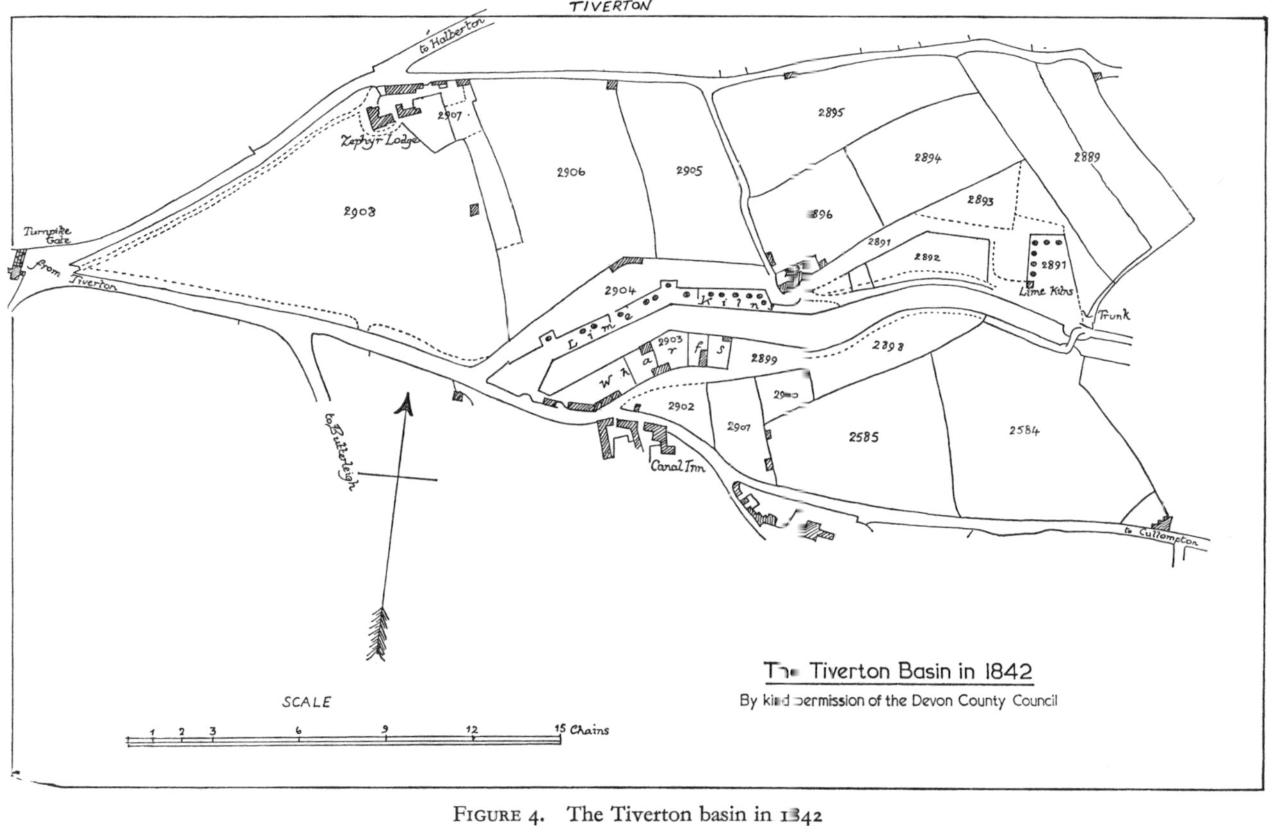

FIGURE 4. The Tiverton basin in 1842

steady rise to £4,926 in 1844. (For income from tolls 1816–54 see Appendix 3). Tiverton basin was the main scene of activity. Here in 1841 wharves and kilns were let to Henry Dunsford, who renewed his lease and increased his holding to 3 wharves and limekilns on the north side, to James Talbot (who also rented Waytown wharf, Holcombe Rogus, near which lime-kilns were probably built at this time) and to J. R. Chave, a lime merchant of Burlescombe who rented a lime wharf with kilns and a garden on the north side assigned to him for 5 years in 1839 at a rent of some £55 pa. Wharves for coal were let to John Richards, to John Kingsbury, and jointly to William and Charles Goodland of Taunton and John Quant of Tiverton who were granted a 7-year lease in 1842 in respect of two wharves on the south side.[1] The wharves at Halberton, Sampford Peverell and Fossend (Burlescombe) were not let, but Payton wharf near Wellington was rented by Thomas Elworthy for £3 a year.

But, although full advantage of the new facility was no doubt taken by traders in the town of Tiverton and the surrounding district for carriage of such goods as were normally transported by inland navigation, even at this time, when there was no real competition, the revenue was far below the earlier hopeful estimates. In these years and those following practically the whole of the trade was westward from the lower to the upper levels of the canal, chiefly from Taunton to Wellington and Tiverton and from Holcombe Rogus to Tiverton, the back carriage of trade from Tiverton or Wellington eastward to Taunton being of only very small amount. Besides this the various expenses of the company, burdened as it was with heavy debts, were high, as the accounts for the year from 1 June 1839 to 31 May 1840, taken for example, will show. (See Appendix 4.) During the 1840s the company repaid various amounts of debt and in 1840–1 spent £668 on providing a new parsonage house at Sampford Peverell—long overdue—as it was required to do by the Act of 1811.

THE RAILWAY ADVANCE

Furthermore, even before the canal was completed, a threat impended from the projected Bristol & Exeter Railway, which was soon making its inexorable advance south-westwards. Work on the railway, which was authorised by an Act of 1836, began soon afterwards; by 1842 the line had reached Taunton, by 1843 it was open as far as Beam Bridge near the Somerset–Devon border, and in 1844, following construction of the Whiteball tunnel beneath the county boundary, it was completed to Exeter. And even more shattering to the canal's prospects than the potential competition of the main railway itself was the almost total subjugation foreshadowed by the proposed branch line from it to Tiverton, which was completed in 1848.

Powerless though it was to stem the advance and finding no means to save its situation, the canal company sought to protect its rights where it could. At one point there was some dispute with the railway company over the routing of its line at Trefusis, where the crossing of the waterway coincided with the canal company's lift; an interview took place with I. K. Brunel, the Bristol & Exeter company's engineer, in an endeavour to get the railway company to move the canal lift higher up and the railway lower down,[2] but Brunel apparently conceded little if anything and the railway was carried almost directly over the lift. Then, at a slightly later stage, some disagreement between the two companies temporarily held up the railway's progress, as the *Western Times* of 4 September 1847 disclosed in its report of the Bristol & Exeter Railway Company's meeting three days previously: 'The Works on the Tiverton branch have been delayed by the difficulty of making any reasonable arrangement with the Grand Western Canal Company . . . but the engineer hopes to open the line by the time anticipated at the last half-yearly meeting.' Possibly the difficulty concerned the excavation under the canal of a cutting for the railway near Halberton and the building of an aqueduct to carry the waterway over it, for on 27 November the same newspaper reported that the canal had been closed for some weeks for construction of the aqueduct, which took the form of two brick arches supporting a cast-iron duct 40ft above the railway, with the spaces between the bricks and iron well puddled; £1,200 had been paid to the canal company in compensation for trade lost during the period.

FINANCIAL COMPETITION AND ITS CONSEQUENCES

The coming of the railway was also a matter of much concern to the Bridgwater & Taunton Canal Company, which in 1841 had, at great cost, completed construction of a mile-long extension of its waterway to the River Parrett below Bridgwater, with a dock at the junction, where canal craft, including tub boats of the Grand Western, could load directly from seagoing vessels. The opening in 1842 of the railway to Taunton and the immediate competition it produced was devastating to the heavily-committed Bridgwater & Taunton company. The company immediately lowered its tolls and in 1844, when competition was intensified by the opening of a coal tramway between Dunball wharf (farther down the Parrett) and the railway, negotiated agreements on toll charges with the Grand Western Company, to which it agreed to pay ½d per ton on coal carried on the Bridgwater & Taunton Canal for every mile up to 14½ that it also travelled on the Grand Western. But the following year, after trying unsuccessfully to turn itself into a railway system, the Bridgwater & Taunton Canal Company became so bankrupt that it ignored its agreement with the Grand Western Company, which was claiming the sum of £403. Relations between the two companies became strained. The Grand Western was suffering its own effects from the railway intrusion which had been immediate—from the peak amount reached in 1844 the receipts from tolls had fallen sharply, with the declining amounts of the following three years averaging only half the 1844 figures—and no doubt saw its only salvation

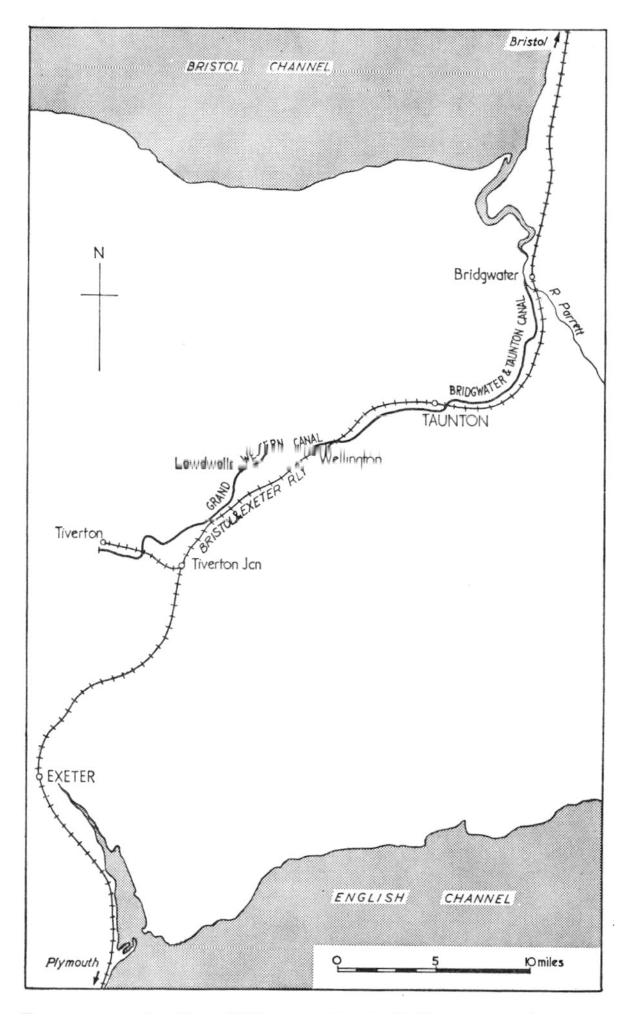

FIGURE 5. The Grand Western Canal, Bridgwater & Taunton Canal, and the Bristol & Exeter Railway, 1848

in some form of agreement with the railway authorities, of which a subsequent fluctuation in the rates received per ton in different years—apparent from comparison of the annual toll income figure and those for tonnages carried (Appendix 5)—were an outcome.

On 8 June 1848 it was reported[3] that an extension coal wharf was being built on the banks of the Grand Western Canal at Taunton and that it was proposed that coal should be brought by railway from Bridgwater and then taken by canal to Tiverton; this, it was said, would save time, but would cut out the Bridgwater & Taunton Canal. A report of 24 June[4] stated that the Bristol & Exeter Company and the Grand Western Company had agreed that the railway should carry all the coal and the canal all the lime, and that there had been local complaints of monopoly since the opening of the branch railway to Tiverton (on 12 June). By July, however, coal was said to be back at its original price at Tiverton 'thanks to some public-spirited individuals'.[5]

Following the completion of the railway to Tiverton the traders on the canal at once opened yards at Tiverton station and used the railway for nearly all their goods coming from Taunton and beyond. Very little traffic was left to the canal apart from lime and stone from Holcombe Rogus and Burlescombe to Tiverton (and the latter of these was seriously diminished) and coal brought off the Bridgwater & Taunton Canal and delivered to intermediate wharves east of Tiverton. In 1851 the railway bargained with the canal company to keep prices high, but this resulted in a greater fall in receipts for the Grand Western Company which in 1852 broke with the railway and renewed its alliance with the Bridgwater & Taunton Canal Company in order to cut tolls and fight the competition. The carriage through to Tiverton of coal brought on the Bridgwater & Taunton Canal to Taunton (where the railway siding by the canal was no longer in use) was resumed, following which the railway company retaliated by promptly lowering its rates. From then until 1854, in order to enable the traders on the canal to sell at Tiverton at the same prices at which the railway delivered, and so to retain a minimum of through traffic on the canal, tolls were drastically reduced to ¼d per mile; even so, the trade on the canal could not be retained and in the last year or so of the period coal was carried over the canal to Tiverton free of any toll whatever. This of course kept down the rates of carriage charged by the railway, which had to bring in its coal at an unremunerative rate or even at a considerable loss, and produced a situation which must have been much to the benefit and liking of Tivertonians.

With the serious reduction in its income caused by the opening of the railway in 1844 the Grand Western Canal Company was unable to pay the annual interest charges on its mortgage debts of £14,750, and in 1853 this interest was consequently 9 years in arrears. For the years 1851, 1852 and 1853 the income of the canal company was not sufficient to defray the current expenses of wages and salaries, repairs and rates and taxes, and a deficiency arose. The situation had obviously become desperate and the company, having no other source of income, acknowledged the hopelessness of trying to maintain coal and other traffic on the canal at a loss in competition with the railway company, which had other resources for meeting any such deficiency. Decision became a matter of urgency when, in 1853, it was discovered that the Bridgwater & Taunton Canal Company had opened a negotiation with the Bristol & Exeter Railway Company for the lease or sale of its canal to the railway. This, it was seen with consternation, would have the effect of closing the Grand Western Canal to all through traffic coming via the Bridgwater & Taunton and, if carried out without any arrangement by the Grand Western Company would, to the company's detriment, undoubtedly close the contest immediately and for all time. The Grand Western Company would be absolutely in the hands of the railway company which, by raising its own tolls and those on the Bridgwater & Taunton Canal to the maximum parliamentary rate, could render the Grand Western's situation hopeless.

Under the circumstances the Grand Western Company, like the Bridgwater & Taunton, decided to offer its canal to the railway company for lease or sale, since it was realised that to keep it open any longer as a separate entity in competition with the railway and to meet current expenses was impossible.

Such a proposal, it was expected, would however be of some financial advantage to the railway company which would receive a direct source of revenue from the continuing lime and other stone traffic from Holcombe Rogus to Tiverton, and, more important, the indirect benefit of being able to charge a remunerative rate for coal transport to Tiverton. The latter was doubtless of some inducement since it was said that the canal company had been causing the railway company a loss of about £6,000 a year by the low rates at which the railway company was compelled to carry coal to Tiverton to enable them to undercut traders on the canal wharf.

Early in October 1853 Henry John Smith, who had superintended and managed the canal since 1845 and had additionally been appointed the company's secretary in 1852, went to Bristol with the authority of the canal company's committee of management and met three members of the Bristol & Exeter Railway board, to whom he put the company's proposals. He found that the railway board was not disposed to purchase the canal but was willing to take a lease. Terms for this were agreed and at the same time the Bridgwater & Taunton Canal committee was asked to give its official consent to the waiving of an agreement made between the canal companies on 29 March 1852 by which either was to give a month's notice before alteration of tolls. This having been obtained, Smith, at the request of the railway company, agreed that the tolls on all goods (excepting stone) passing over the Grand Western Canal on or after 11 October 1853 should be raised to the maximum parliamentary rates. At the same time the railway company made an arrangement with the Bridgwater & Taunton Canal Company by which the rates of toll on that canal would be raised on 11 October to the maximum parliamentary rates in respect of traffic passing to and from the Grand Western, and a new scale of increased rates of tolls for the Bridgwater & Taunton Canal and for the Bristol & Exeter Railway was issued to come into effect on the same date.

The terms of the agreement were contained in a resolution

by the Grand Western Canal committee of 20 October 1853, moved by the chairman, Henry Dunsford, seconded by Mr Hanson and passed with only one dissent, which read:

> That the Committee do now approve and sanction the arrangement then made by the Secretary and Superintendent that the Grand Western Canal Company shall grant a lease in perpetuity to the Bristol and Exeter Railway Company of the Canal Works Lands and Buildings net rents tolls and profits of the canal for the net rent or sum of £2,000 per annum payable quarterly clear of all deduction. The Grand Western Canal Company to undertake to keep the Canal and Works in working repair and that an arrangement be made with Mr Smith for that purpose as suggested by the Bristol and Exeter Railway Company.
>
> The lease to contain a clause giving to the Bristol and Exeter Railway Company power to purchase the canal works lands buildings etc., at any time within 21 years at a sum to be agreed upon between the two companies.
>
> The assent of the Committee being subject to the said arrangement with the Bristol and Exeter Railway Company being approved and confirmed by the Proprietors at a General Special Meeting to be called to consider the same.

It was agreed that H. J. Smith (the secretary and superintendent) should be paid £600 a year (out of the £2,000 rent) for keeping the canal in working repair, with the use of the company's boats free of charge. Finally, on a proposal moved by Mr Hanson and seconded by Mr N. V. Heygate, it was unanimously agreed that the mortgage bond holders, who could hardly expect to derive benefits from the letting of the canal to the full extent of their interests, should be approached so that a just arrangement could be made with them, enabling the committee to make its recommendations regarding the agreement with the railway company to the proprietors.

At the meeting of the company held the following day approval was given to the main proposals of the lease and the chairman was requested to communicate with the mortgage bond holders as had been suggested. A suitable agreement with the latter was of importance because of the difficulty foreseen in getting the assent of the shareholders to the terms of a lease which would leave nothing available to them, for if the 10 years' arrears of interest were to be paid to the mortgagees and their future rate of interest continued at 5 per cent no surplus would have reached the shareholders and probably they would not in this case have confirmed the agreement. Therefore it was fortunate that an arrangement was made by which the holders of mortgages magnanimously consented to give up all claims for the arrears of their interest and to reduce the future rate of interest on their mortgages from 5 to 3 per cent. After full legal attention had been given to the conditions and terms of the agreement for lease and sale and in getting the draft of the Deed, and ultimate agreement on them reached by both companies, a special meeting of Grand Western Canal shareholders was held on 10 October 1854, when formal assent was given to the arrangement and the deeds settled and completed, the handing over being retrospectively dated to 11 October 1853.

DIVIDENDS AND DETERIORATION

In 1854, nearly forty years after the opening of the canal, the company paid its first dividend to shareholders, amounting to the trifling sum of 4s (20p) per £100 share, equivalent to one fifth of 1 per cent, which was repeated in the subsequent remaining years of the company's existence. This was all that was available from the £2,000 income, since payment of the agreed £600 to Smith for repairs, the covering of office expenses and the payment of the reduced 3 per cent interest charges to mortgagees (amounting to £442.50 pa) all had priority, and was only possible in any amount at all through the greatest economy of management. It was a paltry return and a sad outcome indeed for the early promoters' high hopes of the original ambitious canal scheme.

In the decade which followed the handing over to the

railway company trade on the canal decreased even further and became confined almost entirely to a dwindling traffic in lime and road stone on the upper level, from Holcombe Rogus and Westleigh to Tiverton. That on the lower level amounted to virtually nothing; a few boats carrying stone were passed through occasionally, which kept the lifts in some degree workable, but the boats suitable for such passage had become ruinous, the machinery of the lifts and plane old and precarious and the wharves deserted. In general the Taunton–Lowdwells stretch, though still essentially maintained, presented an appearance of impending deterioration and decay. For the canal company there was no hope whatever of any improvement in the situation: the rental it received was almost fully committed in payments which could not be reduced, and any possibility of an increase in the minute dividend was out of the question. Indeed, how long the dividend could be continued was in doubt, for the company was only too aware that in the event of an accident or casualty on the canal there would be a deficiency to meet the payment of any damages which might arise.

SEEKING A BILL

After ten years, having faithfully carried out its part of the agreement—to little benefit—the canal company felt that the time had arrived when legislation should be sought. During this period, in fact, the matter had been suggested on various occasions but the railway company had not so far acceded, maintaining that the proper time did not appear to have yet come. The postponement, it was said, was due to the 'jealous feeling' prevalent in parliamentary committees and held by the Board of Trade that such transactions, involving amalgamation of rival carrying companies, meant loss to the public of benefit arising from competitive rates of carriage. At this stage, however, it was admitted that an absolute sale of the canal to the railway company at suitable terms would improve the position of both parties; the railway company would be relieved of the responsibility and expense of the lease, would profit by the difference between the interest on the purchase money and the rent it was charged, and would save in salaries to redundant officials. The canal company, on the other hand, would be freed from the obligation of keeping in repair a length of navigation of which the greater part was disused and deteriorating, and would be enabled to pay its mortgages and wind up and close the accounts, dissolving the company and dividing any balance between the shareholders.

Correspondence passed and interviews took place between some of the directors of the Bristol & Exeter Railway and the chairman of the committee of the canal company, following which agreement was reached. The terms were embodied in the following minute of the Board of Directors of the Bristol & Exeter Railway Company adopted on 4 November 1863, when they were approved and confirmed:

> Mr Heygate, Chairman of the Grand Western Canal Company and Mr Henry John Smith, their Secretary, attended the Board of the Bristol and Exeter Railway Company on the 28th October 1863, and stated the readiness of the Canal Company to sell to the Railway Company the Canal and all and every the lands cottages materials steam engines machinery fixed and movable plant of every description chattels and all other property and premises of the Canal Company in the Counties of Somerset and Devon with their and every of their appurtenances if terms of sale and purchase could be agreed on.
>
> A negotiation ensued which resulted in an agreement for sale and purchase at the price of £30,000 provided Parliamentary sanction could be procured to such sale and purchase. It was also agreed that application for such Parliamentary sanction should be made in the next session of Parliament, the expenses of such application to be borne by the Railway Company if an Act be obtained, and if an Act be not obtained the expenses of the application to be borne equally by the two Companies.
>
> The sum of £30,000 to cover all claims, outgoings and contingencies whatsoever which are to be discharged by the Canal Company up to the settlement of the purchase.

Page 125 (*above*) The former plane-keepers' cottages at Wellisford; (*below*) the building which formerly housed the steam engine of the Wellisford inclined plane.

Page 126 (*above*) Rock bridge and wharf, Halberton, probably around 1900, with stone heaped on the wharf; (*below*) the scene at Rock as it is today.

Following the agreement the necessary parliamentary notices of an application for a Bill were given jointly by the two companies and the Grand Western Canal and Bristol & Exeter Railway Bill was prepared and settled between their solicitors. Its passage through the House of Lords early the next year was delayed, however, due to queries raised on some of the clauses by Lord Redesdale and by opposition from a group of petitioners.

CONTENDING WITH OPPOSITION

The matters raised concerned mainly questions of public policy involved in the sale of the canal and its partial closure and abandonment, also of the public or private interests of other parties that might be affected by its closing as a navigation, its management when in the hands of the railway company and the disposition of the land of the length which it was proposed should be stopped up and disused. Consequently the Grand Western Canal Company was obliged to prepare a case in support of the Bill which was presented to a committee of the House of Lords on 25 April 1864.

The preamble of the Bill, details of which were being queried, after reference to the earlier Acts, to the making of the canal and to the amount of the existing capital and mortgages, stated that, due to the opening of the railway, the portion of the canal between Taunton and Holcombe Rogus had become almost totally disused and that no public advantage was likely to be derived from its continued maintenance as a navigation. The statement of the case, to which supporting evidence was given by William Partridge of Tiverton, solicitor and principal clerk to the canal company, and by the secretary, H. J. Smith, showed how the revenue received had become insufficient for maintenance and pointed out that the lime and stone traffic on the upper level would not be affected by the closure of the lower length. It also drew attention to the advantage to adjacent landowners of the closure of such a piece of canal, for many years practically disused, and of the lands it occupied being resold, rather than for it to remain useless and stagnant and injuriously affecting the natural drainage of the lands on either side.

Regarding the Bill's third clause concerning the purchase price of £30,000 which Lord Redesdale considered excessive, it was stated that the sum was less than one tenth of the cost; the figure had been arrived at after full discussion and hard bargaining between the companies and was lower than those earlier proposed by the canal company, which might now be considered as having the worse part of the deal. The price in fact represented the equivalent of only 15 years rent by the railway company and a sum based on 20 years might reasonably have been expected by the canal company but for the urgency of its financial state.

Of principal interest to the landowners were clauses in the Bill regulating the disposal of the lands occupied by or forming part of the length of canal to be disused, and it was to points contained in these that the opposition particularly alluded. The period of seven years had been suggested as the time in which the railway company should be required to sell such land, but this was felt to be too long and that two or at the longest three years ought to be sufficient, and an amendment was proposed. There had been considerable difficulty, it was stated, in preparing the particular clause in the Bill which concerned the offering of the canal land for sale. It was originally generally agreed that the people who had sold to the canal company in the first place should have rights of pre-emption, and that only after their refusal to purchase should power of sale to the public be given, but the matter was not entirely straightforward. In most cases the people who sold land to the company in 1830–1 had died or there had otherwise been changes of ownership, and in some places the canal stood on land bought from different people on either side which caused complications, particularly where the former dividing line was no longer easily definable. It was therefore now suggested that the Bill should be amended so as to give the current owners of land adjacent to the canal the right of pre-emption and a further amendment was put forward regarding cases where the land on either side was owned by different people.

The case continued by calling attention to the allegations of the petition which had been lodged against the Bill. The petitioners joining in this opposition were ten in number and were described as 'Owners of land and property adjoining the said Canal'. The first whose signature appeared was Edward Ayshford Sandford of Nynehead Court, whose property abutted on the canal to a far greater extent than that of all the others put together, but he had since withdrawn from the opposition, directing that his name should not be further used in it, which deprived the petition of much of its weight.

The complaints of the other nine appeared to arise mainly out of personal grievances and were demonstrated to be mostly unjustifiable or untenable or to have been already resolved. One statement in the petition related exclusively to a grievance alleged by E. A. Sandford and by Edward Easton of Bradford-on-Tone regarding the uselessness of wharves each had erected on his property. Sandford having withdrawn, only the complaint of Easton had to be considered. The case against it, in which letters which had passed between the parties' solicitors were produced, stated that the land on which Easton's wharf stood was not on the canal's originally planned line, so had not been taken under parliamentary powers but by special arrangement; by this he could only erect a wharf on notice from the canal company that such a wharf was necessary for public use and this had never been given so that he had no right whatever to do so. It appeared that Easton had made the wharf on his land at Allerford, beside the Victory Inn which he also built, about 25 years previously, as a private speculation and chiefly for his own accommodation as a miller. It was not, so far as was known, used by the public. Thus, it was claimed, he was precluded from asking for any compensation on account of the canal being closed and it was further shown that the wharf could be of no value to its owner or the public as it had no trade advantage and had in fact not been used for many years. From the correspondence it was plain that the principal object of the petition, of which Easton had been the promoter and leading agent, was to force the Bristol & Exeter Railway Company to make and allow him the use (at Allerford) of a siding from the railway for the benefit of his trade, on the ground of such accommodation being required in substitution for that afforded by the canal. In spite of the evidence, a clause by which the railway company was required to construct the Allerford siding 'for the Use of the Public' was included in the Act.

THE ACT OF 1864

On 14 July 1864 'An Act for the Sale to the Bristol and Exeter Railway Company of the Undertaking of the Company of Proprietors of the Grand Western Canal, and for the Abandonment of a Portion of the Canal; and for other Purposes' was passed, comprising 42 clauses (27 & 28 Vic, c. 184).

By the Act the canal company was empowered to sell the canal and all its property to the railway company for the agreed £30,000, the sale to take place within six months, and was required to discharge out of the purchase money all mortgage and other debts. If the sale was not completed within six months the railway company was to pay interest on the purchase money at 5 per cent per annum, in consideration of which it should be entitled to the tolls, rates and profits from the canal. Also the company's powers were to be extended to the railway company, as well as its legal obligations.

The existing committee of the canal company, comprising James Heygate, John Remington Mills, Sir Frederick William Heygate Bt., John Oliver Hanson, Brice Pearse, William Unwin Heygate, Charles John Pearse, Henry Wilson, John Barker Huntington and Sir John William Lubbock Bt., were named to continue as such until the dissolution of the company.

The railway company was given power, following the sale and transfer, to abandon the portion of the canal between Firepool bridge at Taunton and Lowdwells lock in Holcombe Rogus parish and was required to sell the lands and works of the disused canal, except any which were required for the railway, within three years. If the company used such land at Taunton lying between the roads to Kingston and Staplegrove for the purpose of the railway, owners of land adjoining should have the right of compensation for any damage caused—a provision which was evidently introduced as a result of points raised by the petitioners.

Regarding the resale of land to adjoining landowners, the Act required that if the land on both sides of the canal was owned by the same person, then the land should first be offered for sale to him, but that if the lands on either side were owned by different people the offer should be made to them in succession, 'in such Order as the Railway Company shall think fit'.

The railway company was required to make compensation in respect of roads crossed or disused by the canal and, as already stated, to provide a siding at Allerford. Liabilities of the Bridgwater & Taunton Canal Company and of the Conservators of the River Tone regarding maintenance of the approaches to the canal were to cease and the portion of the river which the first Act made a part of the canal was to cease to be such.

The transfer of the canal from the Grand Western Canal Company to the Bristol & Exeter Railway Company came into effect on 13 April 1865.

CHAPTER 6

Century of Twilight

HAVING become owner of the Grand Western Canal, which it had acquired not as an investment but because the canal could not exist against the railway's competition, the Bristol & Exeter Railway Company shortly took steps towards closure of the Taunton–Lowdwells length, which was effected in 1867. Machinery was dismantled and removed from the inclined plane and lifts—which soon lost most of their masonry as well—and the greater part of the land which the canal had occupied was sold off and reabsorbed into the adjoining farming country. An example of such a sale, which it may be of interest to quote, remembering the part which the purchaser played in opposition to the 1864 Act, was that made to Edward Easton of Bradford-on-Tone; this was conveyed by a deed dated 6 May 1868 and consisted of a length of approximately 37 chains of the canal from the road bridge at Victory crossing eastward towards Norton Fitzwarren, and involved 3 acres 0 roods 7 perches, the cost of purchase being £91 3s 9d (£91.19).[1] The land of the canal's eastern extremity was retained by the railway company and used largely for the expansion of the goods yards and sidings at Taunton station.

The Tiverton–Lowdwells length was saved from extinction by its continued usefulness for local limestone traffic which survived to a diminishing extent for a further sixty years. (See Appendix 6 for known figures of receipts and expenses.) Stone from quarries in the Canonsleigh area of Westleigh, about half a mile from the canal, was conveyed on a tramway which

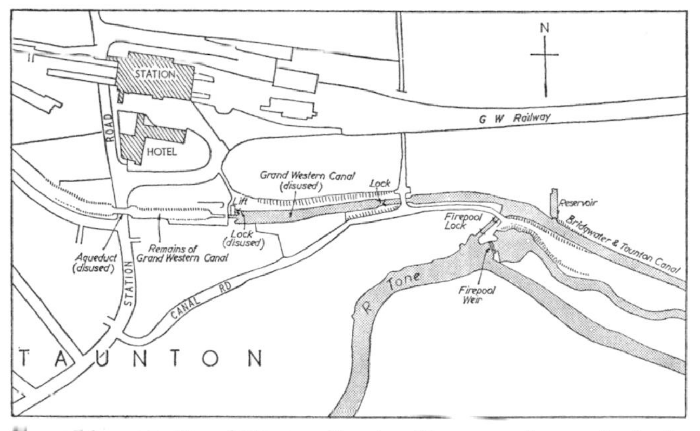

FIGURE 6. The Grand Western Canal at Taunton—the remains in the 1880s, based on the first edition 25in ordnance survey

passed from the quarry through a tunnel under the Westleigh road and then ran on an embankment south-eastwards towards the canal, where the line branched, one section curving northwards to nearby Fossend wharf and the other crossing the canal on a viaduct to terminate at Burlescombe railway station. From a quarry on the Fenacre side stone was carried on another, shorter, tramway a little farther north; this too left the quarry by a tunnel and then went across fields to a short inlet from the canal. Another source of supply, farther north again, was at Whipcott, where stone from a quarry to the west of the canal was also carried by a tramway, and then delivered to the waterside by means of a chute.

Some of the stone was unloaded at various small wharves along the canal for use in roadmaking and for building purposes, while, until the end of the century, quantities were still taken all the way to Tiverton to be burnt in the kilns at the basin. There are local memories of some of these, probably only two, being in use at least up to 1895, leased at this stage to a member of the Cosway family and operated by a man called Berry; boys used to warm their feet by the fire and sometimes roasted potatoes in the embers. The kilns at Tiverton were filled in early this century, in his young days, by Mr William

Punchard, largely with material excavated in making the foundations of houses then being built on Canal Hill, at a charge of 6d a load.

In 1888, by which time the canal was owned by the Great Western Railway Company with which the Bristol & Exeter Railway Company had amalgamated in 1876, the annual tonnage of stone carried on the canal was only a quarter of the average for the years 1846–54. Governmental returns for the years 1888, 1898 and 1905 show the figures as follows:

	Tonnage conveyed	*Revenue* From tolls	From other sources	Total	*Expenditure* Maintenance, management, etc	Profit or Loss
		£	£	£	£	£
1888	4,113	75	122	197	139	58 profit
1898	1,952	36	106 (rents)	142	309	167 loss
1905	5,182	93	141 (rents)	234	343	109 loss

In the year 1904, when the sole traffic was roadstone, it was stated[2] that only two boats were using the canal. These worked as a pair, chained together fore and aft, the leading boat having a pointed bow and carrying 8 tons, the follower being a box boat, carrying 10 tons. Each when loaded drew 1ft 8in of water.

The last man to use the Grand Western Canal for stone traffic was William Elworthy, who lived at Crosslands, Halberton and who used the waterway up to 1924–5 for bringing stone—latterly about 7,000 tons annually—from his Whipcott quarry to a crushing depot which he had at Tiverton road bridge. In these latter years the boats, each carrying 8–10 tons, were worked in one or two sets of three boats each, the leader having a pointed bow and the other two squared. The three boats in a set were linked by chain and wire rope to two horses under the control of two men, one of whom, travelling in the second boat, would steer by means of a large curved stick which he used as a lever against the rear port or starboard of the leading boat, according to requirements. A man called Frank Bell was one who is remembered as having worked in charge of the canal horses in the early 1920s, but the last employed in this job, in his first few years after leaving school, was Mr Charles Pengelly. He can recall being at Crosslands, where the horses were stabled, at 5.30 each morning to take empty boats the 8 miles up the canal to Whipcott to be loaded with stone. The return trip, with a total load of 24 tons, took 2½ hours and the large blocks had to be unloaded by hand. One such journey was made each day in the final years, though in the immediately previous period a second team operated during four months of the year. Mostly the trips were uneventful, though on one occasion one of the heavy shire horses is said to have slipped into the water at Fenacre bridge and drowned.

Small quantities of the stone still continued to be offloaded at small wharves on the route, from where it was taken to roadside landing places for hand cracking by stone-breaking men. But the main bulk was dealt with at the Tiverton road bridge depot, where Mr Elworthy had a steam-driven stone cracker. The machine, which was supported on a big wooden frame and operated by belt and flywheel, was exceedingly noisy and made the atmosphere very dusty as the stone passed through the cylinders before being discharged in different grades of size. The crushed stone was delivered to a wide area around Tiverton—to places as far afield as Cruwys Morchard and Templeton—by a fleet of 23 steam-driven traction engines which were a familiar feature on the district roads, to which their heavy wheels were often detrimental. Perhaps appropriately, the business also owned 14 steam rollers which were hired to local highway authorities for road repair work.

In its later working days the canal played a part in other enterprises. Elderly people can still remember farmers washing their sheep, prior to shearing, in the canal at Fenacre bridge, putting the animals into the water on one side where there were special pens, and bringing them out on the other, their heads being kept above water meanwhile by the use of a special pole. Previous notice had to be given to the railway authorities and a charge was made per score of sheep washed, as it also was for any water extracted from the canal for other purposes.

Another, unusual, feature of life on the canal since early this century was the marketing of the water-lilies which became established along its length, flowering in profusion from June to August. It has been reputed that their growth was initiated by a woman who planted some at Tidcombe long ago but this seems hardly the full story since the slight fall and flow of the canal is in the Tiverton direction. The trade continued until the mid-1960s, sole rights to pick the water-lilies having been held for many years in the form of a lease from the railway company by Joseph Barrie and his family of Sampford Peverell. During the summer the lilies were picked from a boat which was drawn by a slow-walking horse with a long tow-rope, while the pickers—usually three in number—quickly gathered the blooms by severing the stems below water level as the boat moved along. The water-lilies were packed and sent by rail to London, the Midlands and the North where they were used mainly in wreaths. (It is said that on one occasion the railway company confiscated a consignment *en route*, claiming the lilies as its property.) Generally the demand for the flowers was steady but when it fell a boatload sometimes had to be thrown away. Though 11 miles of the canal were formerly covered for this trade the picking area gradually became reduced to about 4 miles due to weed encroachment and other factors and at last, when the work ceased to be economic, Messrs Bert and Joe Barrie finally gave it up.

Recreation, too, was afforded by the waterway in the years of its commercial decline. This was so at least since the 1860s, as evidenced by an extract from an old diary quoted in the *Tiverton Gazette* of 13 January 1948 as follows:

> On Saturday, July 2, 1861, the Tiverton 14th Devon Rifle Volunteer Corps, under Capt. Amory, proceeded with their band to the canal and embarked in boats lent by Colour-Sergt. Robert Davey and proceeded to Canonsleigh for a picnic. They

> sailed as far as the Tiverton road bridge, disembarked, formed fours, marched through Halberton with sloped arms, bayonets fixed. At the New Inn they halted and partook of beer at the Captain's expence. They marched up the hill to Rockway bridge, embarked in the boats again, and through Sampford and disembarked at Canonsleigh.

A collection amounting to £5.50—a large sum for those days—was said to have been made for children at Canonsleigh, and during the water-borne return journey 'it rained all the way'.

A trip along the canal in one of the working boats—after it had been well scrubbed out—with a picnic somewhere along the bank was still a popular holiday activity in the neighbourhood until well into the present century. Sometimes when the water was clear and the weather hot children would take an unofficial bathe. Fishing, too, was enjoyed by an increasing number, from small boys trying their luck with primitive home-made rods to more serious anglers pitting their skill against the waters' numerous quantities and species of coarse fish. There are two life-size paintings in existence of pike caught by Mr Arthur Morton Moncrieff in 1905 and 1906—the larger of them weighed 22lb and was 4ft in length—and records of pike taken up to 28lb in weight. Sizeable perch, rudd and tench as well as very large eels were also frequently caught.

In winter when the weather was very cold and the water froze people flocked to the canal for skating, particularly at the Tiverton basin. Mr W. C. Stone, in his nineties, remembers how, as a boy, he and two of his post office colleagues, named Knight and Snell, skated all the way from Burlescombe to Tiverton, the last one home having to pay the outward journey rail fares for all three. Also due to Mr Stone is the memory of one man falling through the ice into the water and his lady friend blandly calling out to him, 'Keep cool dear'. Perhaps it was this skater's drenched clothes which were remembered by another person as being put near the heat of the limekilns to dry.

The work of maintaining the canal went on through its declining years. The return made to the Board of Trade in 1907 stated that the works were 'generally in good condition' while evidence given to the Royal Commission on Canals and Waterways in the years 1906–9, though stating that the small piece from Lowdwells to Whipcott wharf was 'absolutely impassable for any boat under any circumstances' (presumably being unused, it had been left to deteriorate), indicated that the Grand Western was one of a group of canals which were 'in a very fair working condition' and quite able to cope with a much larger volume of traffic than they were getting, and on which a dredger was kept.

In charge of the canal's maintenance gang from 1913, when he transferred from work on the railway, was John Brewer. Living at Burlescombe he continued in the job until he retired in the early 1920s, and had as his assistants Harry Palk, Jack Colwill and Eli Cook. No dredging was done at this stage, when Elworthy's stone boats were the sole commercial traffic, maintenance being concentrated chiefly on controlling the persistant weed growth (but as far as possible avoiding the water-lilies) and attending to leaks.

Cutting the weeds was done annually by means of a series of knives or saws—latterly, at least, comprising 16 in number—bolted together and suspended on a rope which passed through metal D-pieces attached to the knives. In the earlier days two men (latterly one) stood on either side of the canal holding the rope and pulling forward and back with a reciprocating sawing action to fell the weeds. After so much had been done the cut growth was gathered by the use of a boat in mid-stream which, by ropes attached to bow and stern, was pulled broadside into the bank bringing the weed with it. The weed was then raked out and later generally burnt. Grass and weed along the towpath also had to be kept cut and in winter, where necessary, hedging was attended to and ditches cleaned.

The occurrence of leaks, indicated by a fall in the water level, was mainly in the section between Halberton and Tiverton road bridge, due to fissures in the underlying rock. When a leak was reported the maintenance staff would set off with a boatload of puddle clay and endeavour to find the exact point at which the water was being lost, which was not easy; sometimes, it has been said, a water-lily leaf was dropped into the water as an indicator at the likely spot and would be seen to spin rapidly and be drawn towards the point of escape. Once found, clay would be packed into the fault in order to stem the leak but often much of it just disappeared and remedial work did not last. The trouble was progressive and eventually in the 1930s, after the stone traffic had ceased, the half-mile section in which the leaks occurred was stanked off and left dry in order to eliminate the nuisance and expense. The remaining length from Halberton to Tiverton was thus deprived of its supply of water from the springs at the canal's eastern end and from then on was fed solely by land drainage and surface water.

Sometimes it was an overflow rather than a leak which caused an emergency. One incident of this, reported in the *Tiverton Gazette* in July 1919, was due to the tunnelling of badgers between Halberton and Sampford Peverell, the banks suddenly giving way and land on both sides being flooded; prompt action by the railway employees was said to have contained the flooding but not before the level of water in the canal had dropped by 2ft and several large pike were left stranded. On other occasions heavy rain meant trouble and at the age of 70 John Brewer was known to have walked from Fossend to Lowdwells at 10 o'clock at night to lift the fender regulating the overflow by tunnel to the River Tone. Later, in the 1950s, flooding from the canal at this point had severe effects on the road and property below and in repair work extra drains were laid by the local authority as a future preventive measure.

An annual event in Mr Brewer's time was the railway company's tour of inspection. The inspecting officer, a Mr Sweetland, was said to have been treated with great deference, and during the preceding fortnight considerable preparations were made for his visit. The main boat of the three normally used by the gang was cleaned out and a framework and tarpaulin put over the top, and a special horse, with all brasses cleaned, hired from a local builder. Then, when the inspector arrived at Burlescombe, two men would be sent to carry picnic baskets from the station down to Fossend, from where the entourage would set off for a thorough scrutiny of every detail of the waterway.

John Brewer's immediate successor was a man named Isaac. Another who was soon afterwards to start a period of 43 years service in maintaining the canal, was Mr Charles Middleton, who retired in 1972. The black iron boat which has been in regular use by the maintenance staff of recent years is identical with those which carried stone in the canal's later working days, except that a superstructure of two small cabins has been added. Earlier boats are said to have been of wood covered with metal.

After 1925, when the stone traffic had ceased and the canal's income was negligible, maintenance work was kept to a minimum for reasons of economy. Gradually, as the years passed, the bed of the waterway became silted and weed growth increased. Though despised by some local people as it became gradually more derelict, by others the canal was valued for the recreational attractions it still held. The towpaths provided pleasant walking in a beautiful setting for the elderly, for mothers with prams and courting couples, and, as nature became increasingly dominant, many were drawn to it for the interest of its wild life. Fishing continued, with the rights held by the City of Exeter Anglers' Association, and in hard winters skating on its frozen surface was still enjoyed.

When, on 1 January 1948, nationalisation came into effect, the canal passed from the hands of the Great Western Railway Company to those of the British Transport Commission. In 1962, its obsolescence as a commercial waterway now fully established and acknowledged, the Grand Western Canal was formally closed to navigation, and in January 1964 was put under the ownership of the British Waterways Board.

So ended the hundred years that followed the demise of the Grand Western Canal Company, in which the canal drifted into somnolence and became a half-forgotten feature of the landscape. It had been a century of twilight, but a twilight which was to see a new dawn.

CHAPTER 7

New Life

THE question of what was to happen to the 11-mile length of the Grand Western Canal had, by the 1950s and early 1960s, become a matter for speculation and of controversial discussion. As early as 1951 a suggestion was made at a meeting of Tiverton Trades Council that it would be opportune for the Tiverton Borough Council to purchase a section of the canal for use as a dumping place for hard-core and that at some future date the route might form a by-pass for the town. At the same meeting another thought was expressed, that it would be better if the canal were cleaned out and used as a boating pool. During the ensuing two decades these views represented approximately the dichotomy of opinion as to the canal's future, whether it should be eliminated or preserved.

In 1961 the Redevelopment Committee which reported to the British Transport Commission prior to the Act of 1962, by which the Grand Western Canal (then being managed by British Railways) was legally closed to navigation, found that there was considerable local interest in the preservation of the canal for its amenities. Besides recommending closure, the committee advocated that the canal should be maintained in decent condition and made available for worthwhile projects, such, it was suggested, as transfer to the Devon River Board, as and when required.

The Transport Commission's decision to abandon the canal as a navigable waterway stimulated local discussion. Many people who valued the canal and who could see it as a potential

Page 143 (*above*) The canal near Halberton, about 1900; (*below*) General Sir Hugh Stockwell of the British Waterways Board, and Colonel Eric Palmer, Chairman of Devon County Council, in the weed-cutting boat at the Tiverton basin after the formal handing-over of the canal, May 1971

Page 144 (*left*) The III milestone, at Tiverton road bridge; (*below*) Members of the Canal Management Body, travelling in the maintenance boat drawn by the weed-cutting vessel, making its first water-borne inspection of the canal since improvement work, October 1972. Apart from its added superstructure, the maintenance boat resembles exactly the boats which carried 8–10-ton loads of stone on the canal up to the 1920s.

recreational asset to the area could foresee, in spite of the Redevelopment Committee's recommendations, the pressures that might be exerted upon and by the local authorities for sections of the canal, either in the borough or rural district, to be filled in. As a result, early in 1962 the Mayor of Tiverton and the Chairman of the Tiverton Rural District Council convened a public meeting in Tiverton Town Hall which was attended by people from the town and district who felt that the canal should be preserved if at all possible. At this meeting the Tiverton Canal Preservation Committee was formed, its elected members comprising Mr W. P. Authers as chairman, the Reverend J. R. Dawson-Bowling (chaplain of Blundell's School) as secretary, Mr D. A. Wotton (a master at Tiverton Grammar School and member of the Rural District Council), Mr L. C. White and Mr D. C. Harward. Further meetings were held by the committee and on 24 April 1962 a deputation from it met a representative of British Transport Waterways (the sub-commission controlling the waterways under the British Transport Commission) for a full discussion. Afterwards the sub-commission issued a firm statement, including the points that it had no immediate plans for changes in the character of the canal which could constitute a threat to existing amenities, that if such changes were contemplated in the future the local authorities would be informed and consulted, and that any practical suggestions for the future use of the canal would be welcomed and sympathetically considered.

Having obtained these categorical assurances the committee felt it had fulfilled its immediate task. For the next four years the situation seemed static, the public were unquestioning and the angling association in a spate of renewed activity contributed to an illusion of security.

Then, in the summer of 1966, a crisis arose when news was given that the Tiverton Borough Council was being recommended by its Planning and Development Committee to agree that the portion of the canal within the borough should be filled in by the British Waterways Board and the land used for residential development. The preservation committee's chairman, Mr Authers, himself a borough councillor, reminded the council of the preservation committee's formation in 1962 and of the subsequent meeting with the British Waterways representative. He drew attention to the British Waterways Board's publication *The Facts about the Waterways*, published in December 1965. This stated that to eliminate the 11-mile stretch of the Grand Western Canal between Lowdwells and Tiverton would only slightly reduce the current annual loss of £4,000 on the canal, and, noting that the canal passed through pleasant countryside and had no nuisance problem, pointed out that though receipts amounted to about £200 a year there was an annual deficit of £3,700 which would rise to £4,300 during elimination work. Finally, after listing numerous local organisations and groups of people for whom the canal was a valuable amenity, Mr Authers moved an amendment requesting the council to defer any action regarding the canal until the Tiverton Canal Preservation Committee had had an opportunity to study the problem and to consult the different bodies interested. The council unanimously agreed to the request and the matter was put into abeyance until the council's next meeting on 10 October.

The situation had now become desperate for those intent on saving the canal, and hopes of preserving it appeared to hang by a very thin thread. The committee, now enlarged and strengthened by the appointment of additional members, which included among others the headmasters of Blundell's School and the Tiverton Grammar School, consulted urgently on how best to use the month's reprieve and realised that, once again, much would depend on public opinion which could sway the outcome one way or the other. Various views were being expressed in the local press and at meetings, some extolling the canal's delights and pressing for its retention, others opposed to the idea. Generally, however, it became clear that there was a strong feeling in favour of retaining the canal, but that preservation should be linked with restoration and

TIVERTON CANAL

A Public Meeting

will be held at

TIVERTON TOWN HALL

(by kind permission of His Worship the Mayor)

on WEDNESDAY, 28th SEPTEMBER, 1966
at 7.30 p.m.

All interested in saving our Canal are urged to attend

W. P. AUTHERS *Chairman*
D. C. HARWARD *Secretary*
TIVERTON CANAL PRESERVATION COMMITTEE

Maslands, Printers, 16a Fore St., Tiverton. Tel. 2613

FIGURE 7. Notice of Public Meeting by the Preservation Committee. 28 September 1966

improvement towards its use as an amenity. Having received an encouraging number of promises of support, the preservation committee arranged a public meeting.

Over 200 people attended the meeting which was held on the evening of 28 September at the Tiverton Town Hall; no one present could remember any meeting previously held in the building which had drawn so many. It was addressed first by the preservation committee's chairman who, after outlining the position to date, emphasised the improvement aspect and said it was not known whether there would be a charge on the rates nor if a major appeal would be launched, nor to what extent voluntary labour would be needed. Mr Harward, who had succeeded as secretary, then spoke. He referred to three alternatives facing England's canals as given in the British Waterways Board's publication: complete elimination, reduction to tidy water channels (unsuitable for any boats save light unpowered craft such as canoes) or maintenance in whole or in part for larger vessels. The committee, he said, hoped it would be able to make a case for maintaining the canal under the second category. He then gave the meeting details of the board's general and financial appraisal of the Grand Western Canal—which Mr Authers had previously given to the borough council—and quoted his own recent communications with the British Waterways Board from which, he said, it was clear that the council's current urgency was nothing to do with the Board, but was a matter for the council alone. Finally, to show something of what had been done in canal restoration with the aid of councils in other parts of the country, the secretary instanced the Brecon & Abergavenny Canal, the Stratford-upon-Avon, and canals in Greater London as ones currently being improved.

A four-point resolution was then put to the meeting as follows:

(a) This public meeting strongly objects to the proposal that Tiverton Canal should be filled in and records its determination

to support all efforts made by the Tiverton Canal Preservation Committee to ensure the preservation and improvement of the Canal that it might provide for all time an amenity of enhanced value for fishing, boating, natural history, walking and skating.
(b) The meeting therefore earnestly requests Tiverton Borough Council and the British Waterways Board to defer any action for at least six months to enable a practicable and longterm scheme to be formulated for the Canal and its surroundings.
(c) This meeting requests the Committee to open a Canal Preservation Fund to raise initially not more than £50 to finance the opposition to closure, the administration of such fund to be entrusted to the Chairman, Secretary and Treasurer.
(d) The meeting welcomes the suggestion that a waterside public park might ultimately be provided but recognises that this is a separate project outside the terms of reference of the Canal Preservation Committee but the meeting requests the Committee to support Tiverton Town Council in any such project.

After the resolutions had been proposed and seconded members of the meeting were invited to speak. In response, views were expressed on behalf of various bodies and schools which were represented and also by numerous individuals. The scope of the canal as an amenity and for a wide variety of recreations and its value for level and traffic-free walking—particularly in view of housing developments in the area—was emphasised. Also its considerable educational value for such activities as boat building and handling and, with its extensive flora and fauna, for biological study was described. The sea cadets, their commander said, had used the canal for 11 years for training and offered help in raising money and labour. Such help was also promised by the City of Exeter Angling Association, whose representative stated that the association had spent much time and money restocking the water over recent years and that there were 400 licence-holding anglers in Tiverton. Mr C. F. Unwin, the committee's recently appointed technical adviser, made a statement and gave his opinion that the cost of preparing the 3,000-yard length of the canal within the borough and filling with acceptable material could not be achieved for less than £15,000–£21,000.

In all, 24 people made speeches from the floor of the hall, all of them in favour of retaining the canal, and none against. A vote was taken on the resolutions and, apart from part (c) which was passed after an amendment which proposed raising a sum greater than £50 had been defeated, was carried unanimously. The sum of £40 was raised by an on-the-spot collection and promises given of more. The members of the preservation committee were overwhelmed by the show of public feeling and the success of the meeting, which marked the turning point in the campaign. The unanimity and enthusiasm of the gathering was reminiscent, perhaps, of that earlier meeting, held just a few miles over the hills at Cullompton in 1792, when the idea for the canal's construction was first publicly considered.

The preservation committee now felt it had a firm mandate to go forward to the next stage of the campaign for saving the canal. At the following meeting of the borough council, on 10 October, the committee chairman, Mr Authers, moved an amendment that the council's decision should be deferred for six months to enable all possibilities for preserving the Tiverton stretch of the canal as an amenity to be explored. After an hour's debate the chairman of the council's Planning and Development Committee withdrew the resolution regarding the proposed filling-in on the understanding that any scheme formulated by the preservation committee would not be based on financial help from the council. This assurance was given by Mr Authers and it was unanimously agreed to let the decision rest for the time requested.

On 28 October a deputation from the preservation committee met officials of the British Waterways Board. While prepared to co-operate in trying to find a way in which the canal could be preserved as an amenity, the Board made no promise of financial aid. Relieved up to a point, the committee acknowledged that the problems were considerable, but was still enthusiastic. A start was made in the preparation of an outline scheme to cover several years which, it was foreseen, if acceptable to the British Waterways Board would necessitate launching a public appeal for money and the help of organised teams of voluntary labour. The actual cutting of weed and its removal with the accumulated silt, which could be done by mechanical means, was seen as a relatively minor problem, the size and cost of the operation depending mainly on how far the material had to be taken for disposal and the degree of cooperation of adjoining farmers and landowners. In a repetition of history an approach was made, this time for advice, in the direction of the Kennet & Avon Canal, on which a much larger restoration project had recently been undertaken. Meanwhile it was reported that five local bodies, comprising both young and older people, were making plans for boating and canoeing on the canal if clearance were effected.

In February 1967, following trials and negotiations, the preservation committee was given an estimate of £3,975 for the clearance of weed and silt from the 1½ miles of canal from the Tiverton basin, including the disposal of material which could not be deposited on adjoining land, and was assured that after this work little attention should be required for 20 years. The committee was greatly relieved and, aiming to get the canal cleared for the summer, decided to press for a meeting with the chief engineer of the British Waterways Board, whose contribution would have to be ascertained before a public appeal was launched.

In March the Tiverton Borough Council gave the preservation committee authority to negotiate, and on 6 April a meeting took place at the headquarters of the British Waterways Board in London between the Board's general manager and a deputation from the committee. The deputation returned to Tiverton with a feeling of disappointment; the Board had been sympathetic to the committee's aims and gave permission for the restoration to proceed as and when the money could be raised, but was unable itself to promise financial help, taking the view that since the length of waterway was short and unconnected with a national network the problem was solely a local one. However, the Board wished to consider and take advice on some of the suggestions raised, and promised to communicate its proposals within a week.

On 11 April the Waterways Board's general manager wrote to the committee stating that the board was willing for the canal to be taken over by 'an appropriate Body' with resources and that it wanted the whole 'wet length' from Tiverton to the dry section near Halberton treated as one, rather than just the length within the borough boundary. At this the committee proposed a joint meeting with the Tiverton Borough Council and the Tiverton Rural District Council to consider the situation. When, on 26 May, this took place, with 6 councillors from each authority, the RDC delegation strongly urged that not only the Tiverton stretch but the whole length of canal should be preserved.

The preservation committee met again with the representatives of the two councils on 30 June and the rural district members reiterated their views. Following a discussion it was decided that a long-term policy for the canal was necessary and that consideration should be deferred until a wider survey had been carried out and the possibility of preserving the whole waterway explored. A further delay of just a few weeks was thought justified if a way could be found to embark on a wider scheme which could make the canal a valuable amenity of considerable importance to this part of Devon. At the suggestion of the borough council members, the Dartington Amenity Research Trust (DART), which had been set up earlier in the year to study recreation in the countryside, was invited to make an independent survey and report.

The DART report was completed in October and presented by Mr Michael Dower, the trust's director, on 10 November to a joint meeting of the preservation committee, members of both councils, and officials of the Devon County Council, Devon River Board and the Area Sports Council.

After outlining the canal's history the report noted that, considering the canal was last used for navigation in 1924, apart from the dry section it was still in reasonably good condition, the solidity of its structure partly explaining the high cost of original construction. The original bridges were all in place, no works had been done to restrict the width and headroom and it was still in a single, public ownership. If it were left alone weed growth and silting would continue and the canal would become even less usable and attractive, while the British Waterways Board's annual deficit of £4,000 would continue and probably increase with deterioration. Elimination of the canal would be unpopular and would involve capital costs equivalent to an annual interest charge of £3,700. It was clear that some degree of reclamation was essential in the public interest.

Referring to *The Facts about the Waterways* and the alternative standards of restoration or maintenance it laid down, and taking into consideration the dry middle section of the canal, the report stated that there were three possible degrees of restoration:

1. Two basic channels with a dry section between;
2. Two navigable channels with a dry section between;
3. Complete restoration to one navigable channel

The first possibility, it was pointed out, would provide only limited recreational use and, though it would be cheapest in terms of capital, would involve fairly high maintenance costs. The second, for which increased water supply at the Tiverton end—either by laying a pipe through the dry section or diverting streams—and the clearing of weed growth and silt would be necessary, would leave two unconnected stretches of water which seemed unlikely to attract much public support or income yet would be expensive to create and maintain. The third solution, which would involve reclaiming the dry section and clearing weed and silt throughout to a minimum depth of 2ft 6in, though it would be the most expensive, would certainly be the most attractive for varied public use and could be expected to attract considerable public interest and income.

It was considered that, with development of recreational facilities and improvement of the canal landscape, the canal could form an excellent country park. With population growth and increased leisure and mobility there had been a phenomenal increase in the demand for outdoor recreation, including many activities highly suitable for waterways like the Grand Western Canal. There was growing pressure on the Devon countryside from residents and visitors, calling for new resources for recreation, which the canal, in its beautiful setting, was geographically well-placed to provide. Its wide potential for recreation was envisaged as embracing water pursuits—angling, canoeing, boating, cruising and swimming—towpath uses comprising walking, riding and cycling, special uses such as nature study and local history, and the passive activities of picnicking and camping. The Devon River Authority, it was stated, was keen to see more water brought into use for angling and regarded the Grand Western Canal as having great potential for the development of coarse fishing. The City of Exeter Angling Association of about 3,000 members, which did regular clearance of the canal in the Tiverton–Halberton stretch, would support improvement of the canal fishing by clearance and restocking; the association would not object to the reasonable use of the canal for other purposes such as boating and would prefer full restoration to one navigable channel.

The matter of finance was examined under three headings: capital expenditure, annual maintenance and running costs, and annual income. Estimates of capital expenditure, it was explained, showed considerable variation, depending on the extent to which voluntary labour could be attracted and the amount of work done by contractors. The provisional estimate of capital costs was summarised as on the following page.

While it was clear that considerable capital cost might be involved, particularly if the canal were to be made navigable, the report pointed out that the costs need not fall wholly on the

	Two basic channels £	*Two navigable channels* £	*One navigable channel* £
Clearance of minimum channel	500–3,000	—	—
Clearance of full channel	—	5,000–15,000	5,000–15,000
Water supply for western section	—	500–2,000	—
Reclaiming dry section	—	—	*
Development of fisheries	1,000	3,500	4,000
Park development	*	*	*
Totals, excluding *costs	1,500–4,000	9,000–20,500	9,000–19,000

* Costs which cannot be estimated before further survey.

authority which was to run it. Some help could be expected from other local bodies, such as the river authority and the angling association, a public appeal might raise an appreciable sum and there was almost the certainty of some form of government grant, either directly from the British Waterways Board, or under forthcoming countryside legislation, or under the Physical Training and Recreation Acts.

Maintenance and management costs, to which the British Waterways Board currently allocated a total of about £4,000, would, it was expected, vary from £2,230–£6,540, again dependent mainly on the degree of help forthcoming from volunteers and also on whether or not full navigation facilities were provided.

Regarding income, the report stated that this was currently about £220 a year, including £46 rent from a cottage at the Tiverton end, £2 for fishing rights and the remainder from water supplied and miscellaneous items. A provisional estimate of annual income which might be expected following a reclamation scheme was:

	Two basic channels £	*Two navigable channels* £	*One navigable channel* £
Angling licences	100+	250+	250+
Canoe and boat licences	15	25	50
Motor boat licences	—	110	1,100
Rents	150	150	250
Income from camping etc	25	50	100
Profit on water-buses	—	—	250
Water supply	70	70	70+
Voluntary contributions	10	50	100
	370+	705+	2,170+

N.B. This is the income as it might be in the fourth or fifth year after reclamation work started. Income in the first two years would be a little more than the present income of some £220 p.a.

Comparing the provisional estimates of annual costs and income, it could be seen that the total annual deficit was small at any degree of restoration and that, taking the worst figures for each scheme, the most ambitious (the one navigable channel) was only marginally more expensive than the least ambitious (two basic channels).

Whatever degree of restoration was to be decided upon it would be necessary, the report noted, to consider the nature of the body which should administer it. There seemed to be three choices: the British Waterways Board, the local authority or a voluntary body. The Waterways Board, it was considered, would be unlikely to be able to devote much of its resources to reclamation of the Grand Western Canal in its separation from the cruising network, but it was empowered by the British Transport Commission Act of 1962 to transfer a waterway to any local authority, statutory water undertaking or river board, or to lease or transfer it to a voluntary body. Local authorities, already having general powers to assume or assist such undertakings, were due to acquire further powers in forthcoming legislation, including making grants for 'country parks', for which the reclamation and development of canals would be eligible. They also had skilled staff for engineering, planning and other related work, together with the permanent financial backing of their rateable income. The third possibility, a voluntary administering body, might, it was said, attract voluntary labour and subscriptions and could qualify for grants but might lack the resources to maintain the canal once reclaimed. The most suitable administration for the Grand Western Canal appeared, therefore, to be the local authority, possibly backed by a canal association.

In conclusion the report noted that the borough and rural district councils would wish to discuss matters with interested bodies, the chief of which were: British Waterways Board, Devon County Council, Somerset County Council, Devon River Authority, the Canal Preservation Committee, the City of Exeter Anglers' Association, the South Western and the County Sports Councils, the Devon Trust for Nature Conservation and related bodies, and youth groups. Such discussions would show the measure of support for a restoration scheme. Before definite decisions could be made, however, further surveys were needed to study the engineering aspects and to assess the possible demand for recreation and the need for park development.

The meeting which received the report agreed to appoint a working party consisting of representatives of both councils, and hoped that the Devon River Authority and the county planning authority would assist in the endeavour. At the same time the preservation committee's name was changed to the Grand Western Canal Preservation Committee to indicate its determination to preserve the whole canal.

During the succeeding months the two councils studied the DART report in detail, the Devon River Authority prepared an engineering report and the working party met at intervals. At a meeting held on 16 October 1968 the working party resolved to have experiments undertaken on the canal's dry section and to ask the county planning officer to carry out a survey on the recreational demand and park development. Up to this stage no decisions had been taken regarding the statutory powers under which the local authorities should participate in the maintenance of the canal and the county council had not been formally represented on the working party. However, during 1968 the Countryside Act had come into force which gave county councils power to establish country parks, a form of which could be an inland waterway, and also, in the same year, the British Waterways Board had become empowered under the Transport Act to make an agreement with a local authority for an inland waterway to be maintained by it or taken over.

Early in 1969 the British Waterways Board indicated that it was prepared to hand over the Grand Western Canal to the local authority, together with a sum of money for maintenance, which would be based on the capitalised value of the cost of maintaining the canal over the previous few years. The Clerk to the Devon County Council communicated the information to the working party, stressing that no figure had so far been negotiated and stated that:

> The appropriate committee of the County Council are to recommend the council to approve, in principle, the vesting of the canal in the Devon County Council, subject to a satisfactory financial contribution being agreed with the British Waterways Board and subject also to satisfactory arrangements being reached as to management and finance with the Tiverton Borough and Rural District Councils. They would, of course, also have regard to the interest of the Canal Preservation Society.

The working party, meeting in June, was unanimous in its support of these suggestions and agreed that the council be asked to continue negotiations with the Board with a view to

SAVE THE CANAL

TOW PATH
WALK

Saturday, 18th October, 1969

The Lock Cottage, Greenham - -	10.30 a.m.
Burlescombe (Fossend) Canal Bridge -	11 a.m.
Sampford Peverell Bridge - -	12.45 p.m.
Halberton Road Bridge - -	2 p.m.
West Manley Bridge - - -	2.30 p.m.
Tidcombe Bridge - - -	3 p.m.
Tiverton Basin - - -	3.30 p.m.

3.30 p.m.

CIVIC WELCOME by the MAYOR OF TIVERTON

TIVERTON TOWN BAND SEA CADET DEMONSTRATION

Issued by the Grand Western Canal Preservation Committee

Maslands, Printers, 16a Fore Street, Tiverton. Tel. 2613

FIGURE 8. Notice of Towpath Walk by the Preservation Committee. 18 October 1969

the promotion of the Grand Western Canal as a tourist project.

In July the Devon County Council agreed without debate to its finance committee's recommendation. It was now suggested that British Waterways Board would be willing to hand over the canal with about £25,000, though the Board had stated that such a capital sum would be difficult to find in the current financial year but could be budgeted for in the next if the county council indicated acceptance. The county surveyor, however, contrary to the view of the Board's local engineer, felt that the annual maintenance of the bridges, if capitalised, represented a much larger sum, and on this, as well as on other issues, negotiations continued.

Meanwhile there was still some variation of local opinion regarding the canal's future, and partly because of such feelings, in order to test the measure of existing public interest in the restoration of the canal and also to provide an opportunity for people to discover and appreciate its less familiar reaches, the preservation committee organised a towpath walk along the whole eleven miles of the canal's length.

The walk, which took place on Saturday 18 October 1969, was planned in every detail with the care and precision of a military operation. Invitations were issued in advance to numerous organisations in the district, to individuals and generally to the public, and arrangements were made for committee members to act as 'marshals' at the various joining points. Car parking facilities were worked out, the Blundell's School Signal Section was co-opted to provide communication along the route and the Red Cross for the provision of any necessary first aid. On the morning—when a message of good wishes had been received from a 98-year-old former employee of the Great Western Railway, who had handled canal affairs—about 400 walkers assembled at Lock Cottage, at the Lowdwells end of the canal, and were set on their way by Tiverton's Member of Parliament, Mr Robin Maxwell-Hyslop, who fired a starting gun. On the route another 400 people joined in and when all reached the Tiverton basin the assembly, with others who had gathered there, totalled about 1,200 and included parties from 34 organisations. Though five hours had been allowed for covering the distance some young people arrived in less than three. An 11-year-old boy fell into the water near Rock bridge, others travelled by canoe. A lady of 68 walked the whole length and declared her readiness to do the return journey. At Tiverton, where the event was covered not only by the local press but also by the two television authorities and by a reporter and photographer from *The Times*, the crowd was met by the mayor and by the vice-chairman of the rural council, the town band played, the sea cadets gave a demonstration and refreshments were available. The enthusiasm was tremendous, the occasion, described by the preservation committee's chairman, Mr Authers, as 'a massive manifestation of enthusiasm for the canal to be restored', was almost comparable with those which marked the inauguration of canals in earlier days.

Negotiations between county council and Waterways Board continued during the following months. The project was a sizeable one, with numerous different bodies involved, and it was clear that development depended on the amount of finance available and on the extent of any government grant. Early in 1970 it was rumoured that the council had bought the canal subject to contract but this was denied; in fact, the canal was being given by British Waterways Board, together with a sum for maintenance, and it was on the amount of this sum that discussions were prolonged.

In April news was given that verbal agreement had been reached. The canal was to be a gift, with £30,000. That same month Devon County Council published a report on the Grand Western Canal, prepared by the County Planning Officer, Mr Phipps Turnbull, the proposals of which, following up the DART report, were intended to provide the basis for discussions between the departments towards the proposed country park. It emphasised the canal's suitable recreational situation for the neighbouring populations of Devon and Somerset—which were expected to increase—and also the particular advantage it could provide as a counter-attraction in relieving tourist pressure on Dartmoor, Exmoor and the coastal areas, particularly with the advance of the M5 motorway, which would increase the flow of day visitors and holidaymakers from Bristol and the Midlands. The canal's potential recreational uses were reiterated and details given for possible lines of development, with suggestions for providing special facilities at various points.

In addition to the interest on the sum promised by the Waterways Board it was estimated that a further £10,000 per annum would be needed for maintenance and running expenses. During the summer it was agreed that, to meet this, £2,500 should be contributed each by the borough and rural district councils and the remaining £5,000 by the county council. In the autumn a new management committee for the canal, called the Grand Western Canal Management Body, was set up, comprising one member from the borough council, one from the rural district council, two from the county council and one from the preservation society; Messrs W. P. Authers, N. Jones, D. C. Harward, W. H. Ayre (chairman) and P. S. Day were appointed to fill the respective places, though Mr Jones was replaced by Mr R. M. Britton shortly afterwards. The preservation society opted to remain in existence, agreeing that, while its main objectives of preservation and improvement were in sight, it could still give help and guidance; it also felt it had a mandate to maintain vigilance, an understandable viewpoint after such a strenuous and protracted campaign.

Though proceedings appeared to drag, negotiations prior to the canal's conveyance continued through the winter and involved many lengthy and complicated legal matters. In addition to the important question of how the money was to be paid by the British Waterways Board, these included discussions with owners of land through which the canal had originally been cut, consideration of liability for bridges, walls and wharves, and sorting out such problems as whether British Waterways Board or British Railways owned the aqueduct over the former railway line and which of them was responsible for its maintenance. A large number of infringements, many of them of long standing, had to be patiently examined by the legal department of the county council.

Eventually all was ready for the canal to be handed over and on 5 May in the Tiverton Town Hall the contract, which was to be effective on 24 June 1971, with a cheque for £38,750, were officially passed by General Sir Hugh Stockwell of the British Waterways Board to the Chairman of Devon County Council, Colonel Eric Palmer, who described the negotiations and progress as 'an exercise which demonstrates goodwill, co-operation and common sense'. After the ceremony the meeting adjourned to the canal, where the main personalities of the occasion embarked on a brief voyage across the Tiverton basin.

The money which the Waterways Board had given was put on deposit to provide an 'insurance policy' for future liabilities and not allocated for country park development, an application to the Countryside Commission for the canal to be treated as a country park—which would qualify for a 75 per cent grant towards costs—being rejected. Retaining the services of the three maintenance men who had transferred from British Waterways Board, the management body with its limited financial resources set to work, concentrating initially on main essentials for the basic interests of walking, angling and boating.

One of the first jobs to be done was the clearing of weed growth. For this purpose a French weed-cutting boat with vertical and horizontal cutting bars was purchased by the management body, an investment which proved highly effective along the whole of the watered length. Silt removal, an operation which presented some difficulties due to inconvenience of access and the narrow towpath, was then tackled. With the co-operation of adjoining landowners and the use of three Hy-Mac machines the towpath was opened and levelled so that a dragline and bucket could be brought in to dredge the silt from the canal bed. The dragline was then worked from the towpath, the silt being thrown to the opposite bank; this made the canal very slightly narrower, since some of the silt fell back into it on the far side, but funds were not sufficient for the material to be carted away.

Having cleared the waterway, and at the same time provided a more level and wider towpath, the management body turned its attention to the half-mile dry section. The restoration of this was important not only to give a continuously navigable channel but also to enable the proper flow of water from the springs at the eastern end to the Tiverton length, which since the stanking-off had been fed only by surface water and land drainage. This work was undertaken for the management body by the Devon River Authority and the seriousness of the leaks (caused by severely fissured sandstone bedrock in which clay plugging was never lastingly effective) was demonstrated dramatically at the outset, when 2 million gallons of water pumped into a dammed-off portion of the dry section disappeared in 10 days. The problem has been dealt with by the application of a butyl plastic lining sheet, covered for protection with a coating of clay.

With the exception of the weed-cutting boat the management body decided to ban powered craft from the canal, mainly on account of likely damage to banks and the clay-puddled bed, and also because of their undesirability to the considerable angling interest, but the way has now been made clear for rowing boats, small skiffs and canoes. Both fishing and boating are discouraged, however, on the 2½ miles from Fossend to Lowdwells, which is designated as a nature reserve.

And so the Grand Western Canal, topic of discussion through two centuries, after being in existence for 160 years and lying idle and half-forgotten during the last 50, goes forward with an injection of new life. Almost without exception the scheme of restoration now meets locally with whole-hearted approval. Perhaps the hoped-for dividends, which never materialised financially for the subscribers of the early days, may be reaped increasingly by their successors in the form of recreation and enjoyment, which, in the new age of leisure, the canal has the potential to provide.

CHAPTER 8

Itinerary of the Canal Route

WHETHER glimpsed at intervals by travellers on the A373 road through Sampford Peverell and Halberton to Tiverton, or known more closely by people who have lived in its proximity, the surviving Grand Western Canal, in its serenity and unspoilt beauty, transmits a silent yet irresistible invitation to those who would come and discover its tranquil delights or get to know it better. With its working days far behind and having come safely through a period of threatened annihilation, the canal breathes new life and promises refreshment and recreation for those seeking an escape from the business of modern existence. For the energetic walker there are eleven miles of towpath rich with variety and interest, pleasantly rural throughout, and quiet—save when the occasional sound of blasting near Westleigh reminds one of the quarries which provided the canal with its trade, or at a point near Burlescombe, when main line trains thunder past on the railway which caused the canal's commercial demise. For those who just want to amble along a shorter length—to 'unwind' and observe the wild life, hearing only the piping call of the coot or the squawk of waterhen—there is convenient access at the numerous points where bridges cross the waterway, as well as at each end.

The 11-mile Tiverton–Lowdwells section of the canal is freely accessible to the public throughout its length. The 13½-miles' stretch from Taunton to Lowdwells, which has long been abandoned, is not so easy to follow, but even here much of the distance is marked by public footpaths which provide easy-to-rough walking in most pleasant countryside.

THE TIVERTON–LOWDWELLS SECTION

The broad basin at Tiverton, with its shimmering water and its swans, provides a refreshing sight for those who have climbed the steep hill south-eastwards from the town. Today, the surrounding area is primarily residential, but in the past the immediate vicinity at least would have had a more commercial atmosphere, with boats arriving and unloading, horses and carts coming and going away with loads of lime, and the acrid fumes from the limekilns drifting across the water. Some of the relics of those days have vanished, others remain.

Apart from the edge of the quay, which is paved, the immediate surroundings of the basin are mainly grass-covered. In the canal's working days the south bank of the basin comprised four wharves and included some buildings which are now gone. The quay on the northern bank gave immediate access to the limekilns, which were built backing into the basin's retaining structure, their tops being approximately at canal level a few yards from the water's edge. This was particularly convenient, since coal and limestone could be tipped alternately into the top of a kiln with minimum haulage, and the burnt lime extracted from it in the yard below, ready for dispatch. From a map of 1842 it appears that there were then 14 kilns here; the arches of most of them are still plainly seen at the lower level though their tops on the canal bank have long been filled in. A pair of metal plates set into the quay cover a flight of steps which lead down to a former storage cellar accessible from the lower level. The thatched cottage which stands at the end of the yard, just past the sea cadets' *Training Ship Hermes* was formerly the Lime Kiln public house.

About 250 yards east from its commencement the waterway narrows to the canal's average width of 30ft. Still remaining at a point where it narrows further are the abutments of a former bridge. In the masonry are slots for a stop-gate, for use in cutting off the water from a section of the canal during repairs or when otherwise necessary. A few yards to the north, in the area now covered by modern housing development, there was formerly a block of 6 more limekilns. Continuing eastwards, with the towpath on the northern bank, after nearly a mile the canal makes a sharp left-handed bend, widened on the corner to make the turn easier for boats, and immediately passes under the round-arched Tidcombe bridge. A few yards past this is the I milestone and a short distance farther along the Lower Warnicombe accommodation bridge. This is similar in construction to the other accommodation bridges on the canal, consisting of metal (formerly wood) on stone abutments, and, as with other bridges, also having stop-gate grooves on either side.

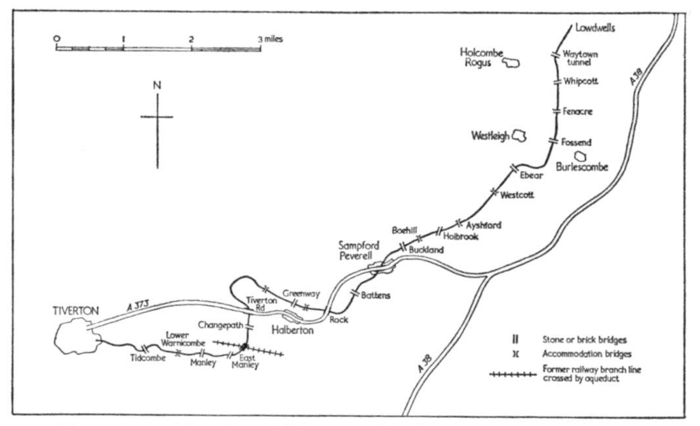

FIGURE 9. The Grand Western Canal. Tiverton–Lowdwells

Manley bridge, the next, which carries a minor road, is similar in form to that at Tidcombe, stone-built and substantial. On the following stretch the II milestone is reached and the canal passes under a brick-built round-arched bridge carrying a lane from East Manley Farm; the waterway widens on the south bank into what may have been a lay-by or private wharf and then curves to the north. Here it passes by aqueduct across the former branch railway line from Tiverton Junction to Tiverton, which was constructed 1847–8 (see p 116) and closed for passenger service in 1964 and for goods in 1967. Though the structure was built with twin arches the railway line always remained single.

Beyond the aqueduct the canal continues northwards, past the remains of a small wharf on the right bank, where in the past stone to be hand-cracked for use in roadmaking was unloaded, and then passes under 'Change-path' (formerly Back Lane) bridge (of stone) which carries a minor road from Halberton Lower Town. The bridge is so-called because at this point the towpath changes from the left to the right bank, and the horse had to use the bridge to cross from one side to the other. Within a very short distance the canal passes beneath another, known as Tiverton road bridge, also of stone, which carries the A373; immediately past it the III milestone can be seen on the right bank and the site of a wharf and Elworthy's former stone-crushing depot on the left, the waterway taking a sharp bend westwards. From here the canal forms a wide loop (sometimes known as the swan's neck) by which it curves around clockwise to approach Halberton in a south-easterly direction.

Near the IV milestone is a bridge carrying a path which leads to Sellake Farm, and beyond this, as far as Greenway bridge, is the half-mile section which, in the 1930s, due to persistent leakage, was stanked off so that it became dry. The stone-built Greenway bridge carries a minor road from Halberton to Uplowman; beyond it the canal runs eastwards, north of Halberton village, and, after a short length of cutting, is crossed by an accommodation bridge (formerly a swing bridge) carrying a minor track. A cottage on the canal's south side, just a short way along, was built, it is believed, at about the same time as the canal, and it is said that the man who lived in it had the daily duty of inspecting a certain distance of the canal in each direction and checking the stability of the embankments. Between here and the point known as Rock the canal is carried on an impressive 50ft high embankment, the construction of which must have largely accounted for the remark by Rennie in his 1813 report that the Halberton lot had been 'the most difficult and expensive piece of Work of its kind on the Canal'. At the east end of the embankment stands the V milestone. On the northern side of the canal both to the immediate west and east of Rock bridge, which is here in sight, are disused sandstone quarries from which it appears likely that the dressed stone of Rock bridge itself and of others between here and Tiverton was obtained. The scene here, on the western side of Rock bridge, is today perhaps the most attractive of any on the canal's length; the site of the former wharf on the north bank is clearly defined, and the garden of the house beside it is in springtime bright with daffodils and forsythia, while in summer the waters of the canal itself support the leaves and cool blooms of water-lilies.

Before the canal's construction the line of the road was some yards to the west of Rock bridge, which now carries the A373 for the second of its three crossings of the waterway. Rock (formerly Rockway) House, situated on the east side of the road just above the bridge, was built at about the same time as the canal or shortly after for John Twisden, the retired naval captain who superintended the later stage of the canal construction to Taunton, notably after the departure of James Green. It is possible that Twisden, who apparently had a large family of seven daughters, had a private wharf here with an underground passage for conveying coal and other goods to the cavernous cellars of the house; a tunnel still exists for a short distance below the house but is blocked for any remaining length it may have run to the canal. Past the bridge the canal continues south-eastwards before swinging around to run in a north-easterly direction, along an exposed stretch where canal workers have long known the bite of the east wind blowing across the valley from beyond the Blackdown Hills. There is a likely lay-by here, on the western bank. Battens (formerly Batlins) bridge, constructed of red brick with stone coping and carrying a minor road, crosses the canal on this section following which, past the VI milestone, the village of Sampford Peverell is approached. The site of the former wharf here is plainly seen on the left bank, close to the present Wharf Guest House. The canal virtually bisects the village, to which its construction must have brought considerable upheaval. In 1811 Sampford Peverell was the scene of the navvies' riot of tragic consequence (see p 38). Soon afterwards, following the Act of the same year, numerous dwellings were bought by the canal company and demolished to make way for the canal, which also necessitated slight re-routing of the through road. The bridge which carried this road (the third crossing of the waterway by the A373) is another of brick construction, with coping of stone.

Shortly after passing beneath the bridge the canal bends to the east and is slightly widened on the left bank into what may have been a lay-by; an overflow weir discharges here into a culvert which carries water under the canal. Also on the left is the former rectory, which lost part of its garden to the canal; the present rectory, which the canal company had to provide (see pp 39, 63, 114) stands above it on the other side of the road. The next bridge, carrying a minor road and also brick-built, is Buckland bridge; just beyond it, where the towpath is carried briefly on a footbridge, a low stone arch in the adjacent wall marks the point where at one time water was taken for use in the swimming pool of a former boys' home. On the north bank is the widening of another likely lay-by, and within a short distance the VII milestone appears beside the towpath and the metal-structured Boehill bridge, carrying the lane to Boehill Farm, is reached. Past this, just to the north of the canal, are the overgrown remains of a clay-pit from which much puddling material for the canal was obtained.

Holbrook bridge, of brick except for its stone-coped parapets, is the next, carrying the minor road to Holcombe Rogus. The canal continues north-eastwards and in about ½-mile passes beneath another, metal, bridge near Ayshford

Court. Beyond this there is a straight 1-mile stretch on which are the VIII milestone and Westcott bridge, another typical accommodation bridge, which leads to Westcott Farm. At the end of the straight stretch is the stone-built Ebear bridge by which the minor road to Westleigh crosses the waterway.

Past Ebear bridge the canal bends to the east and is carried on an embankment for another, shorter, straight stretch. Almost half-way along it is the IX milestone and, just past this, one of the numerous streams which cross beneath the waterway is conveyed through the embankment, here reinforced with stone, in twin culverts. The next bend, to the left, is almost a right angle, and marks the point where the Tiverton branch of the canal joins the originally intended main line which, if completed, would have extended from here to the Exe estuary. Immediately past the bend there is a widening on the left bank and at its edge a weir and sluice by which surplus water is discharged to drain to the last-mentioned stream.

For the remainder of its route to Lowdwells the course of the canal runs almost due northwards. Within a short distance it is crossed by a bridge of steel spans on stone piers which formerly carried a tramway from the quarries at Westleigh to Burlescombe railway station; in the early days before crossing the canal the tramway curved to run alongside it on the west bank. Fossend bridge, of stone construction and carrying the road from Burlescombe to Westleigh, is immediately ahead, with the site of the former wharf just before it on the left. From here the Westleigh limestone quarries—far more extensive now than when just the Canonsleigh section was worked, but still retaining some old limekilns—are a dominating feature to the west. Past Fossend bridge and about ¼-mile along is an inlet from the canal's west bank to a former minor stone-loading wharf; this was connected to another part of the quarry area by a small tramway (stone blocks of which are still *in situ*) which left the quarry by passing through a tunnel (now blocked) under the road and then ran across the fields to the canal.

The next bridge is Fenacre bridge, of similar construction to that at Fossend, which carries another minor road leading to Westleigh. From here the canal runs in a deep cutting, another factor—made necessary by Rennie's decision to lower the summit level—which inflated the cost of construction, and in ½-mile reaches Whipcott, where it passes under another stone bridge. There are quarries here on either side, that to the west, when worked by William Elworthy, being the last to supply stone carried on the canal. The wharf to which stone from this quarry was brought by tramway and let down by chute was a few yards south of the bridge and is now overgrown. There was another wharf north of the bridge and the cottages here were formerly a smithy. Two old limekilns remain as part of farm buildings along the lane, there are remnants of others at the entrance to the quarry on the canal's east side, and about ⅓-mile farther along the canal, close to another small disused quarry, is a further group of five limekilns on the west bank.

Immediately beyond this the canal narrows and enters the short Waytown tunnel which carries it under the Holcombe Rogus–Wellington road beside Beacon Hill. Here the horses were again detached from the boats and taken up to cross the road—and also the canal, since the towpath reverts to the western bank. The boats were hauled through the tunnel by using a chain, a short length of which can still be seen attached to the wall inside the circular stone-arched south portal. The springs which form the canal's main supply of water are situated at either end of the tunnel, deep down in the canal bed.

North of Waytown tunnel the canal bends north-eastwards for the last ⅓-mile to Lowdwells. On the right bank, just beyond the tunnel, is the site of a small wharf with the wharf house above it. Farther along, just before the end is reached, there is a sluice on the left bank; surplus water overflows here into another, smaller tunnel, about 50ft in length—undoubtedly the 'heading or tunnel' Rennie referred to as being nearly completed in his report of 1813 (p 50)—and discharges through its stone arched portal at the roadside below. From here the water is carried away in piped drains and within a short distance joins the River Tone. The canal itself, with a cottage on the left bank, terminates in the remains of the lock by which this section connected with the narrower, descending length to Taunton.

THE TAUNTON–LOWDWELLS SECTION

The point where the Grand Western Canal connected with the Bridgwater & Taunton Canal lies to the immediate south-east of the present Taunton railway station. The site can be reached from Station Road, south of the station, by turning into Priory Bridge Road and then immediately left into Canal Road. Continuing for about ¼-mile, passing the railway goods yard on the left and the cattle market on the right, one comes to Firepool lock, where the Bridgwater & Taunton Canal, turning to the south, passes beneath the road and joins the River Tone. The junction of the two canals was on the north side of the minor road bridge.

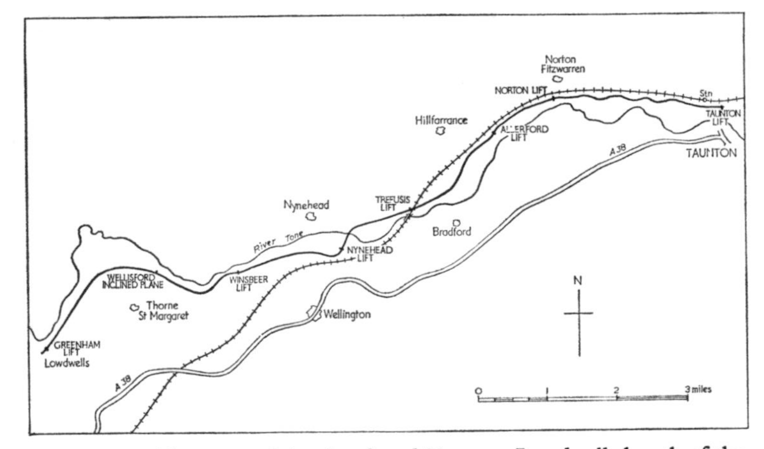

FIGURE 10. The route of the abandoned Taunton–Lowdwells length of the Grand Western Canal, showing positions of the lifts and inclined plane

The former course of the Grand Western is here quite indiscernible, being completely obscured by the railway development. It ran westward along the present boundary of the goods yard, with the regulating lock about 70yds, and the first (Taunton) lift, which was of 23½ft, just over 200 yds from the commencement. The position of the former aqueduct which carried the canal over the road leading to Rowbarton and Kingston St Mary (now Station Road) is marked by the present bridge which carries the railway loop line.

From here to Fairwater bridge the canal route is now occupied by the loop line and Fairwater sidings, but past the bridge the route becomes traceable. A track which passes to the south of allotment gardens and close to a row of brick cottages marks the line of the canal which continues as a public footpath through arable fields to Silk Mill, where it comes close to the south side of the level crossing of the main railway line. Silk Mill bridge, which carries the Bishop's Hull–Staplegrove road over the former canal, has a slightly curved, almost square arch; it is constructed of reddish-hued stone and brick, with visible iron ribs, and, partially covered with ivy, is still in reasonably good condition.

It is not possible to follow the canal route for the next half mile or so of its course, but it can be rejoined by following a public footpath leading westwards from the road which runs north of Bishop's Hull village, after the road crosses the River Tone and just before a second bridge which carries it over the Halse Water or Norton Brook. The path leads almost due westwards towards a cattle shed, past which the line of the canal shortly becomes apparent as a grassy depression with irises growing in it and in some places containing water. The towpath is also clearly discernible. North of Longaller the canal route is crossed by a footbridge and a short distance farther on, south of Norton Fitzwarren station, is the site of the 12½ft Norton lift. There are no actual remains, the lower surroundings being very overgrown, though its position can be determined by the sudden rise in level. From here the route—rather less clear but still mainly detectable—continues through the fields to the Victory Inn, the approach to which is in close alignment with the railway. At the Victory Inn the remains of the wharf and of related buildings can be seen and across the minor road the canal route becomes visible again, partly containing water, to the site of the 19ft Allerford lift. The perpendicular rise of this, though overgrown, is plainly seen, and nearby is the former lift-keeper's cottage, recently renovated.

The route continues for a stretch on an embankment but then, through farmland, becomes less easily traceable. The next accessible point is just north of Tone Green, where the minor Bradford–Hillfarrence road crosses the former canal by a stone-built bridge which has a narrow circular arch lined with brick. Here, where the canal still contains water, there was a wharf. Water continues through the cuttings of the following ½-mile stretch which runs directly south-westwards, passing beneath the Nynehead road at Trefusis Farm by a wider, flattish arched stone bridge to the site of the 38½ft Trefusis lift. Here the railway dominates the scene, its embankment including a brick bridge, partly filled in, which carried the water of the canal. The lift, of which little remains, was immediately north of the bridge, which was built practically on top of it. Close beside it is the former lift-keeper's cottage, still inhabited.

For the next mile, in which the route crosses the minor road near Clavenger Farm, the canal remains are almost obliterated, but an interesting relic, which the canal approached on a slight embankment, is the aqueduct which carried it over the River Tone. The stonework of the wide, graceful arch of this slender 30ft-span structure is still in remarkably good condition and, though the parapet walls are crumbling, it can still be crossed, when the iron trough which contained the water will be seen.

The route, continuing as a public footpath, runs south-westwards through a field on a slight embankment, with a roughish stone culvert carrying a small stream beneath it as it comes within sight of Poole brickworks; shortly after it reaches the site of a wharf, immediately beside the Nynehead road, where the wharfinger's cottage and a shed remain.

Across the road the canal route continues south-westwards for a short distance through an area of felled trees, to reach the former Nynehead lift with a rise of 24ft. This site of all the seven shows the greatest amount of remains, and is therefore the most interesting. Although no machinery survives and the central pier is gone, much of the walling of the side and end walls of the chamber still stands and the arches in the retaining walls, as well as holes in which the framework was secured, are plainly apparent.

Directly after the lift the route of the canal, carried now on a high wooded embankment, crosses by a fine ornamental ashlar arch the former driveway to Nynehead Court. The iron trough of the waterway is still in position. About 50 yards to the south another arch, which carries the railway, can be seen; this, even more impressive than the one carrying the canal, is said to have formerly incorporated a gatehouse.

From here the line of the canal, after a fairly sharp turn, runs almost due westwards to Tonedale, north of the town of Wellington. For much of this distance, through woods, it is an easily discernible depression, but in the latter part it has been levelled out and incorporated into fields, and a footbridge of iron and masonry which formerly crossed it has been recently demolished.

At Tonedale the canal crosses the Wellington–Milverton road just south of the River Tone, where there was a private wharf, and, itself forming the footpath, crosses the mill leat by a short aqueduct. Continuing, and entering fields, it approaches the site of the 18ft Winsbeer lift. Of this there is little to be seen apart from undulations of the ground, while the former lift-keeper's cottage, later enlarged and used as a farmhouse, is now in ruins. Past this the route comes very close to the south bank of the River Tone, though high above it and supported by a retaining wall, while on the south side of the former canal are sandstone quarries from which much stone used in the canal's construction was undoubtedly taken. The canal route, which is here not always clearly defined, continues in fairly close company with the river and follows the same general direction to Harpford bridge, substantially built of stone without iron ribs, which carries the road to Langford Budville.

Beyond Harpford bridge the canal route continues parallel with the river for a short distance but then swings away at the foot of the slope of the Wellisford inclined plane. The line of this, which raised the boats a height of 81ft over a distance of 440ft in a gradient of about 1 in 5½, can be determined in a grass field, with the area of the basin in which the boats waited for ascent still apparent at the foot, containing water and surrounded by trees. On the south side of the basin can be seen the mouth of the adit which drained the bucket well; it is 2ft 9in wide and 3ft high to the top of its round arch. At the top are two well-kept stone built cottages, formerly occupied by those who worked the inclined plane, and at right angles to them the building which housed the steam engine, now used as a store. The former bucket well is now filled in.

From the top of the plane the route continues due westwards, until it crosses the minor road running north from Thorne St Margaret, and then bears to the south-west. The next road crosses it at a sharp corner due east of the ancient Cothay Barton, after which the canal route and road are closely aligned for a distance to Elworthy Farm. Here the route of the canal passes close to the farm before forming a further discernible straight stretch across fields to Greenham, where it is crossed by the road immediately south-east of the village. From here the route approaches in a short distance the site of the 42ft Greenham lift, at the foot of which it is crossed by another stone bridge. Though there is much overgrowth the dramatic rise of the lift is quite obvious, with the keeper's cottage, still inhabited, at the top. Beyond the lift the canal route passes from Somerset into Devon and is carried on an embankment for the short remaining distance to Lowdwells, where the abutments of the aqueduct which carried it across the narrow road can be seen. At Lowdwells a weed-filled depression marks the approach to the junction of this and the earlier-constructed summit level, which is retained. Here, still visible, are remains of the 45ft long lock which raised the boats the 3½ft to the summit level, consisting of the stone edge wall showing grooves in which the lock gates were held.

Notes

NOTES TO CHAPTER 1 (*pages* 11–30)

1. Harrowby Papers (Tiverton), 24 October 1768.
2. *Exeter Flying Post*, 4 October 1792.
3. *Exeter Flying Post*, 31 January 1793.
4. Harding *Historical Memoirs of Tiverton*, I, 1845.
5. *Exeter Flying Post*, 4 July 1793.
6. *Exeter Flying Post*, 13 August 1793.
7. Exeter City Record Office.
8. Somerset County Record Office, D.P.8.
9. Exeter City Record Office.
10. *Exeter Flying Post*, 16 July 1795.
11. Relevant documents are in the Exeter City Record Office.
12. *Exeter Flying Post*, 28 April 1796.
13. *Exeter Flying Post*, 6 July 1797.

NOTES TO CHAPTER 2 (*pages* 31–55)

1. *Exeter Flying Post*, 1 February 1810.
2. *Exeter Flying Post*, 1 March 1810.
3. *An Authentic Description of the Kennet & Avon Canal*, 1811.
4. *Exeter Flying Post*, 29 March 1810.
5. Manuscript copy of minutes of meeting 12 April 1810. Somerset County Record Office, DD/X/CK.
6. Somerset County Record Office, DD/X/CK.
7. Report of the Committee of Management, 24 June 1811.
8. Counsel's brief in support of the Grand Western Canal and Bristol & Exeter Railway Bill, 1864.
9. Extract from *Taunton Courier* in the *Annual Register* for 1811 (27 April) and *Exeter Flying Post*, 2 May 1811.
10. From this detail of the Act, and also according to the *Plan of the Proposed variation from the line of the Grand Western Canal in the Branch to Tiverton* in Devon County Record Office (D.P.15), it appears that the canal was intended to continue beyond the eventual Tiverton basin, following around the contour south of the town, Little Silver being the area near the confluence of the Rivers Lowman and Exe.
11. Devon County Record Office, Q.S.A10/354.

12. *Exeter Flying Post*, 13 June 1811 and in subsequent issues until 22 August.
13. *Exeter Flying Post*, 24 December 1812.
14. Somerset County Record Office, DD/X/CK.
15. Somerset County Record Office, DD/X/CK.
16. This proposed aqueduct was in fact apparently not constructed.
17. Devon County Record Office, 74B/MB 11/(Burrow & Co).
18. Great Western Railway Records.
19. Harding *Historical Memoirs of Tiverton*, I, 1845.
20. Counsel's brief in support of the Grand Western Canal and Bristol & Exeter Railway Bill, 1864.

NOTES TO CHAPTER 3 (*pages* 56–75)

1. *Exeter Flying Post.*
2. Devon County Record Office, 74B/MB 12/(Burrow & Co).
3. Counsel's brief in support of the Grand Western Canal and Bristol & Exeter Railway Bill, 1864.
4. Newspaper cutting in Exeter Public Library.
5. Devon County Record Office.
6. Devon County Record Office.
7. Hadfield, Charles. 'James Green as Canal Engineer', *Journal of Transport History*, I, no 1 (May 1953).
8. Harris, Helen and Ellis, Monica. *The Bude Canal*, 1972.
9. See my Appendix 13 'Inclined Planes, an Assessment of Fulton's Contribution', *The Bude Canal.*
10. Great Western Railway Records.
11. Counsel's brief in support of the Grand Western Canal and Bristol & Exeter Railway Bill, 1864.
12. Devon County Record Office, 74B/MB 12/(Burrow & Co).
13. Great Western Railway Records.

NOTES TO CHAPTER 4 (*pages* 76–111)

1. *Exeter & Plymouth Gazette*, April 1830.
2. Report of committee, 27 June 1832. Devon C.R.O., 74B/MB 12.
3. Great Western Railway Records.
4. Devon County Record Office, 74B/MB 12/(Burrow & Co).
5. Dickinson, H. W., *Robert Fulton, Engineer and Artist*, 1912.
6. Hadfield, Charles. *The Canals of South and South East England*, 1969, and Kenneth R. Clew, *The Somersetshire Coal Canal and Railways*, 1970.
7. Hadfield, Charles. *The Canals of the West Midlands* 2nd ed, 1969.
8. Hadfield, Charles. *The Canals of South West England*, 1967 and Kenneth R. Clew, *The Dorset and Somerset Canal*, 1971.
9. Hadfield, Charles. *The Canals of the West Midlands*, 2nd ed, 1969.
10. Hadfield, Charles. *The Canals of the East Midlands*, 2nd ed, 1970.
11. Green, James. 'Description of the Perpendicular Lifts for passing boats from one level of Canal to another, as erected on the Grand Western Canal', *Transactions* of the Institution of Civil Engineers, 2 (1838). Obtainable as a reprint from the Waterways Museum, Stoke Bruerne, Northants.
12. *Taunton Courier*, 21 August 1833.

13. Great Western Railway Records.
14. Great Western Railway Records.
15. Great Western Railway Records.

NOTES TO CHAPTER 5 (*pages* 112–131)

1. Details of leases in Devon County Record Office.
2. Devon County Record Office.
3. *Exeter Flying Post.*
4. *Exeter & Plymouth Gazette.*
5. *Exeter & Plymouth Gazette*, 15 July 1848.
   Also Counsel's brief in support on the Grand Western Canal and Bristol & Exeter Railway Bill, 1864.

NOTES TO CHAPTER 6 (*pages* 132–141)

1. Somerset County Record Office.
2. de Salis, Henry Rodolph. *Bradshaw's Canals & Navigable Rivers of England* (1904).

NOTES ON CHAPTER 8 (*pages* 166–179)

Research material in the collection of the Somerset Education Museum and Art Service includes the following:

A report on the investigation of the Nynehead Lift

A report by Basil Brass on his investigation of the Wellisford inclined plane.

Author's Notes and Acknowledgements

It would not have been possible for me to write this book without the kind help of many people, all of whom I wish to thank most gratefully.

Mr Charles Hadfield, whose own book, *The Canals of South West England*, provided me with my basic knowledge of the Grand Western Canal's history and also with the incentive and guide for further research, was the person primarily responsible for my embarking on the work. His kindly encouragement and help and very generous loan to me of his own research material have made the task a most pleasant and absorbing occupation. Mr David St John Thomas also encouraged me to undertake the project and kindly lent me his file of relevant material.

I have had the best co-operation possible from the various archivists and their staffs whom I have consulted. The Devon County Archivist, Mr P. A. Kennedy and his assistant Mrs Cameron, the Somerset County Archivist, Mr Ivor P. Collis and his assistant Mr D. R. Jones, and the Senior Archivist of the Exeter City Record Office, Mrs Rowe, have all been unstinting in making available documents for my investigations and in helping me over facts, and I acknowledge with gratitude the access I have been allowed to material contained in the deposit of Messrs Burrow & Co in the Devon Record Office. I would also express thanks for the help I have received from Mr R. J. Hannam of the British Transport Historical Records

Office, Mr H. C. Richardson, Librarian of the Institution of Civil Engineers (with whose permission I am allowed to reproduce the drawings of James Green's lifts on pp 89, 90 and 107), the staff of the Map Room of the British Museum, the Committee of Tiverton Museum, and Mr R. L. Hutchings, Curator of the Waterways Museum, Stoke Bruerne, who has made available numerous items of relevant interest for my use.

I have been grateful also to the respective librarians of the Devon County, Exeter City and Taunton Borough Libraries in which are preserved much of the newspaper material referred to, and in addition express my thanks to the editor of the *Tiverton Gazette* and the editor of the *Devon & Somerset News* for their interest and co-operation.

My thanks are extended to all those people who have corresponded or talked with me about the canal, or who have allowed me to make investigations on their property, of whom I would specially mention Mr L. Brewer, Mr A. Elworthy, Mr C. H. Grabham, Mr V. H. May, Mr C. Middleton, Mr C. Pengelly, Mrs Tooze and Mrs B. Williams. And I have been very glad to be able to have the opinion of Mr Frank Hawtin, Organiser, Somerset Education Museum and Art Service, regarding my investigations of the remains of the canal's abandoned Taunton–Lowdwells length, in which he has carried out archaeological work of his own.

Regarding the canal's recent history and development I have been extremely grateful for the very kind and willing co-operation given me by Mr W. P. Authers and Mr D. C. Harward, both of whom have lent me their files relating to the campaign towards the canal's preservation, in which both have played a very great part. And I gratefully acknowledge the help given me by the Estates Surveyor of Devon County Council, Mr R. G. Pengelly, and by Mr J. S. Neason, Principal Land Agent in that department.

I thank most sincerely Mr Dudley Weatherley for his kindness and skill in preparing for me the drawing of the boat on the canal on p 81 and also Mr Lee Flaws for making the copies of the old photographs on pp 126 and 143 and overcoming certain technical difficulties involved, and for preparing others.

Lastly, I am grateful to my family who have shown their accustomed tolerance during my times of Grand Western preoccupation. Particularly I thank my father and mother, Mr and Mrs J. R. H. Warren, for their many kindnesses during my research visits, and my husband, Desmond Harris, for his never-failing interest, encouragement and support.

Appendix 1

The lengths, rises and falls visualised by Rennie in 1794.

	Length				Rise		Fall	
	Miles	*Furlongs*	*Chains*	*Links*	*Ft.*	*In.*	*Ft.*	*In.*
The line from Topsham to Taunton								
From low water in R. Exe near Topsham to departure of branch to Cullompton	16	1	–	–	200	5	–	–
From branch to Cullompton to near Pugham, departure of Tiverton branch	5	7	–	–	102	–	–	–
From Pugham to west end of summit level	–	2	2	40	24	–	–	–
Summit level	3	1	1	20	Level			
From east end of summit level to near Wellington	4	5	9	40	–	–	152	–
To proposed basin at Taunton	6	3	6	60	–	–	113	1
Total	36	4	9	60	326	5	265	1

	Length			Fall	
	Miles	*Furlongs*	*Chains*	*Feet*	*Inches*
Branch to Cullompton					
From point of departure from main line to basin at Cullompton	1	73	–	30	–
Branch to Tiverton					
From point of departure from main line to Pugham	–	1	1	22	–
From near Pugham to West Manley	5	5	8	Level	
From near West Manley to basin at Tiverton	1	7		56	–
Total	7	5	9	78	–
Feeders					
From the first reservoir above Culmstock to the summit level near Pugham	6	7	2	–	–
From the large reservoir on the Tone to the summit level near Loudwell Mills	1	5	7	–	–

Appendix 2

Names of Subscribers to the Grand Western Canal as listed (and spelt) in the Act of 1796.

Robert Abraham Jnr.
Robert Abraham of Crediton
Gideon Acland
Ann Acland
John Baker
William Brown
Edward Boyce sen.
Edward Boyce jnr.
William Bussel *Clerk*
Samuel Bansill
John Blackmore
John Brutton *Clerk*
Henry Brutton
Robert Baker
William Besley
Joseph Brutton
William Browne
Joseph Cooke
Richard Hall Clarke
John Creswell
Edward Crosse
William Aldridge Cockey *Clerk*
Francis Colman
Henry Carew
James Coleridge
John Cole
Henry Crosse
John Culme
Thomas Edward Clarke *Clerk*
John Chave
John Bradford Copleston *Clerk*
John Clarke *Clerk*
Nicholas Dennys
Henry Dunsford
Martin Dunsford
Richard Down
Francis Rose Drewe
John Rose Drewe
Edward Drewe *Clerk*
Herman Drewe *Clerk*
William Drewe
George Dunsford
Samuel Davy
William Davy
Martha Davy
Sarah Davy
Elizabeth Davy
James Dunsford
Richard Rose Drewe
Edward Eagles
John Ellicott
Jonathan Evans
Ezekiel Abraham Ezekiel
John Fowler
Ralph Fowler
John Finnemore

John Fergussone
John Fortescue
William Farrant
Christopher Flood
Francis Fairbank
George Follett
James Garrett
John Gervis
Henry Gervis
William Baring Gould
Edmund Granger
Richard Graves
Thomas Gray
Sir Alexander Hamilton
John Holmes
Thomas Heathfield
William Hurley
Richard Hurley
John Hogg
Joseph Hogg
John Hanson
John Holman
William Ingram
John Jones
Blomer Ireland
William Kennaway
John Kestle
William Keats
William Lewis
Richard Lardner
James Lardner
John Ley *Clerk*
Samuel Lichigaray
Samuel Lott
John Fownes Luttrell
Francis Fownes Luttrell
Charles Leigh
Mary Mayne
Samuel Milford
John Miles
Humphry Mills
Henry Mills
William Haycraft
John Merivale
Henry Melhuish
Samuel Morgan
Gilbert Nevill Neyle
Thomas Newte
John Nott *Clerk*
John Nott
John Norman
John Needs
George Owen
Sir Robert Palk *Baronet*
Thomas Prowse
Robert Pell
Claud Pell
Albert Pell
Sydenham Peppin
Thomas Preston
William Quick
Thomas Rennel *Clerk*
—— Rogerson
William Roberts
Francis Rogers
Mark Sanders
Hugh Skinner
George Skinner
John Smith
George Shute
Daniel Salter
John Seale
Stephen Shute
Nicholas Spencer *Clerk*
Richard Smith
Henry Smith
John Bacon Sweeting
George Short
John Salter
William Turner
Joshua Toulmin
Thomas Tanner *Clerk*
Thomas Tayler
Ann Bodington Vye

John Watts
Hannah Wilkes
William Walker *Clerk*
John Whitter
Nicholas Were
James White
George Waymouth
Mark Westron
Joseph Watson
John Wood
John Welsford
Henry Waymouth
John Wedgewood
Josiah Wedgewood
Thomas Winslow
Joseph Woolmer
Palk Welland *Clerk*
The Right Honourable Sir
George Yonge *Baronet*

Appendix 3

Toll Receipts, 1816–54

	£	s	d
1816	574	6	3
1817	Not known		
1818	556	13	7
1819	627	5	2
1820	752	10	7
1821	632	1	1
1822	Not known		
1823	401	16	7
1824	381	1	5
1825	575	3	1
1826	700	19	8
1827	739	0	7
1828	623	15	2
1829	705	2	8
1830	Not known		
1831	756	19	11
1832	918	1	8
1833	839	18	10
1834	968	9	2
1835	970	11	0

	£	s	d
1836	1,163	19	11
1837	1,822	19	9
1838	2,754	4	0
1839	3,494	0	0
1840	3,456	19	3
1841	3,631	19	7
1842	4,114	15	1
1843	4,768	15	11
1844	4,925	10	8
1845	2,819	6	6
1846	2,302	12	1
1847	2,291	15	11
1848	1,735	8	6
1849	2,351	18	4
1850	1,660	8	3
1851	890	5	2
1852	974	19	11
1853	734	9	10
1854	956	1	11

Appendix 4

Accounts of Receipts and Expenditure from 1 June 1839 to 31 May 1840.

Received

	£	s	d
Rents received	370	19	5
Tonnage	3,456	19	3
Several boats sold for	76	10	0
Lime Stone, an Iron Tank etc sold for	45	1	4
Thomas Newman for half the expense of repair Allerford Lift	8	18	2
For land sold	1,894	8	1
	£5,852	16	3

Paid

		£	s	d	£	s	d
Rates, taxes and tithes					155	10	1
Rents					69	18	7
Salaries		328	4	3			
Wages, lift and plane keepers		516	2	6	844	6	9
Repairs.	Contract upper Level	312	0	9			
	Houses and boats	3	18	2			
	Lower level, earth work	196	1	9			
	Machinery	332	4	6½			
	Materials	228	13	5½	1,072	18	8
Expenses.	Passing trade in barrows	67	14	3			
	Coal for the engine and Smith's oil etc	99	3	10	166	18	1
	Expenses of sub-committee, stationery etc	24	9	0			
	Sundries for prosecuting Wotton	6	1	10			
	Captain Twisden to London	10	0	0	40	10	10
New Works.	Tiverton Wharf	557	5	5			
	Shaft and chain for plane	18	10	6			
	Stop gates at Taunton and Welsford	50	15	3			
	Cleaning the canal	54	6	4			
	Waste weir Trefusis	3	4	7			
	Boundary walls Welsford plane	19	3	10			
	Stoodham Bridge Nynehead	54	4	6	777	10	5
Purchase of	Land tax, Tiverton	188	6	0			
land etc and	Land	1	10	0	189	16	0
damages.	Pd. Mrs Bird for Damage				5	0	0
Captain Twisden	Balance of account				118	14	9
Rent of office 1½ year to Christmas					60	0	0
Arrear of Salary to Secretary					275	0	0
Wages to Messenger, Postages, carriage of parcels, Stationery, Expenses at London Tavern for room etc					54	15	0
Interest paid on Bonds					1,000	0	0
Balance of repayments for sums advanced in preceding years					390	0	11
Balance of Cash at Sir J. Lubbock & Co		425	7	10			
Ditto	Dunsford & Co	206	8	4	631	16	2
					£5,852	16	3

Appendix 5

Tonnages carried 1846–54 (to nearest ton)

Year ended 31 May	*Coals*	*Culm*	*Lime-stone*	*Build-ing stone*	*Road stone and gravel*	*Brick yard goods*	*Timber and Iron*	*General merch-andise*	*Lime, manure, etc*	*Totals (nearest ton)*
1847	12,693	7,234	21,230	3,096	2,528	462	558	2,431	92	50,326
1848	13,056	9,072	22,989	1,887	6,309	690	1,153	1,158	195	56,512
1849	2,675	7,468	20,815	2,797	1,186	443	23	1,869	140	37,417½
1850	3,281	7,537	14,341	1,981	730	504	93	1,308	—	29,776
1851	3,467	5,960	14,708	1,925	1,777	411	207	1,517	75	30,047
1852	3,345	7,229	14,970	4,142	1,144	717	168	1,509	326	33,552
1853	4,632	6,013	16,519	3,889	2,622	1,193	750	1,544	191	37,354
1854	2,962	3,160	19,229	4,189	3,173	857	578	1,501	103	35,754

Appendix 6

Receipts and Expenses for the years 1869–72 and 1877–92
(from GWR records)

	Receipts			*Expenses*		
	£	s	d	£	s	d
1869	357	15	10	130	17	5
1870	374	7	11	176	19	3
1871	365	17	5	131	13	2
1872	430	18	0	152	9	7

	Receipts	*Expenses*	*Profit*	*Loss*
	£	£	£	£
1877	287	264	23	–
1878	258	233	25	–
1879	260	190	70	–
1880	272	188	84	–
1881	262	459	–	197
1882	274	266	8	–
1883	303	175	128	–
1884	271	159	112	–
1885	206	162	44	–
1886	220	147	73	–
1887	177	250	–	73
1888	198	139	59	–
1889	137	184	–	47
1890	218	150	68	–
1891	199	171	28	–
1892	174	216	–	42

Appendix 7

The following bye-laws, included in those issued by authority of a special shareholders' meeting of 28 November 1838, are especially interesting:

II That if any Boatman, having the care or command of any Barge, Boat, or line of Boats, shall suffer the same to be navigated on this Canal without being himself, or having some male adult person, duly qualified on Board to steer such Barge, Boat, or line of Boats, and also a male adult person attending to drive the Horse drawing the same; or if any person shall navigate one Boat only without a sufficient rudder guiding the same on this Canal, every such Boatman, or other person, shall forfeit and pay for every such Offence any sum not exceeding *Five Pounds*.

III That no person shall navigate on this Canal two or more Barges fastened together, whether loaded or unloaded, neither shall any more than four Boats be fastened together, to pass in one line, whether loaded or unloaded, under the penalty of not exceeding *Forty Shillings*.

XVI That no Boat, Barge or other Vessel shall be permitted to navigate on this Canal armed with Iron or other Metal projecting either laterally or in front, so as to be liable to injure or damage any other Vessel, or any of the Works belonging to this Navigation; and no Boat will be allowed to pass the Plane and Lifts whose extreme length exceeds 25 Feet, or whose extreme breadth exceeds 6 feet 3 inches, or whose Floor is not level and smooth, turning up at each extremity four inches at least in the last foot of the length, and whose sides are not rounded off at the four corners at least 4 inches in the foot, so as to allow them to pass the Caissons without damaging the Works.

XVII No Boat shall be allowed to pass the Plane or any of the Lifts after dark, nor when it shall freeze, and the Ice be of the thickness of half an inch or more; nor of which the Master or Conductor shall be intoxicated, or who shall be in violent altercation with any other person. And the Owner or Conductor of any Boat endeavouring or threatening to force an entrance into the Caissons or Locks of any of the Lifts, or the Plane, without the permission of the Company's Agent there stationed, shall forfeit for each Offence any sum not exceeding *Ten Pounds*.

XX That no less number of Boats than four shall pass the Lock at Loudwell without permission in writing, given by one of the Company's Agents, under a penalty of not exceeding *Twenty Shillings* for every offence.

XXVII The said Company of Proprietors will not be answerable or accountable in damages for any injury or damage done to any Boat or Barge, or to the Contents or any part thereof of either of them by reason or in consequence of any accident happening upon the passing of any Boat or Barge up, into, upon, or through any Lock, Lift, or Inclined Plane, upon or belonging to the said Canal.

Appendix 8

Summary of Facts about the Grand Western Canal

	Length	*Level or Rise*	*Type of boat accommodated*	*Opened*	*Fate*
Tiverton to Lowdwells	11 miles	Level	Barges	1814	Used commercially until 1924–5 Closed to navigation 1962 Now preserved for recreation
Taunton to Lowdwells	13 miles	Rise of 262ft (see below)	Tub boats 8–10 tons	1838	Abandoned 1864

Locks, Lifts and Inclined Plane (On Taunton–Lowdwells length only)

Locks Number — 2 (excluding those which formed parts of lifts)
Situations — Taunton, stop-lock
Lowdwells, rise 3½ft

Lifts Number — 7
Situations — Taunton, rise 23½ft
Norton ,, 12½ft
Allerford ,, 19ft
Trefusis ,, 38½ft
Nynehead ,, 24ft
Winsbeer ,, 18ft
Greenham ,, 42ft

Inclined Plane Number — 1
Situation — Wellisford, rise 81ft, length 440ft, gradient approx 1 in 5½.

Index

References to illustrations are printed in italics

N.B. This index does not include separate references to the names of subscribers listed in the Act of 1796, which are given in Appendix 2, pp 188–90.